PLOUGHING, POLITICS
and FELLOWSHIP

PLOUGHING, POLITICS and FELLOWSHIP

by

Alfred Hall, MBE

For Frank

Every Good Wish.

Alfred Hall

April 2002.

ISBN 0 9539271 0 5

Published by
Alfred Hall

Printed by
Dixon Printing Co. Ltd, Kendal, Cumbria.

To my two helpful daughters,
Margaret Ann and Ingrid Elizabeth.

Ox Power

Horse Power

Steam Power

Diesel Power

Man (and Woman) Power

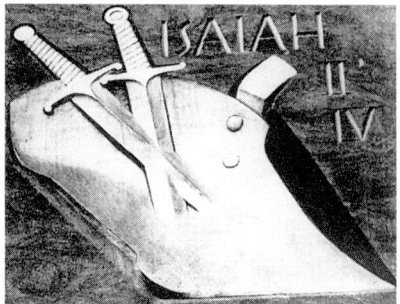

Soul Power

CONTENTS

INTRODUCTION

"Ploughing, Politics and Fellowship" asserts from experience that amongst people involved in the basic husbandry of land, international rural fellowship is neither a myth nor a figment. It is real, active and peacefully powerful.

Since time immemorial the labour of those who peacefully conserve and produce food from many varieties of soils on earth has been symbolised by the sign of that fundamental, ubiquitous and long-lasting tool – the plough.

This narrative recalls how, shortly following the end of the Second World War, when internal combustion engines were rapidly replacing the power of draught horses on farms, and aeroplanes began to offer speedy and frequent transport between countries, a local ploughing match became the catalyst that widened the horizons for ploughmen around the world, inspired national ploughing competitions in many countries and established a world organisation for the "Olympics" of ploughmanship.

The idea was born in the year 1947, cultivated for six years and launched as a viable proposition in the year 1952. Since 1953 World Ploughing Contests have been held around the world in a different country each year and prove that international rural fellowship does really exist, even where lives may be dominated by regimes of essentially different political dogma. Bridges are built over frontiers and "the brotherhood of man transcends the sovereignty of nations".

Preface
ONE WORLD

Poets and writers have picturesquely described the ploughman as "ploughing his lonely furrow". Away in the distance, across a broad landscape he may appear a remote and lonely figure as to and fro over the field he carefully guides his plough turning the earth into a new pattern of colour. He is not alone nor is he lonely. Many thousands of his fellow ploughmen all around the arable world are also ploughing their "lonely" furrows. Though they may be separated in distance they are, nevertheless, working closely together. To any ploughman the arable world is one big field all of which requires the same primary cultivation process in order to grow food-producing crops. Whether the piece of land he ploughs happens to be in Italy or Finland, Canada or New Zealand, the British Isles, Africa or elsewhere his mind and heart are intent upon the same objective. He wants nothing more than to cultivate his field in peace, concentrating on how best he can work with Nature.

He has to plough at the right time, at the right depth, at the right speed and in the best way to combat weeds and make an appropriate, good bed of soil for the seeds he will plant from which food crops will grow.

Ploughing is fundamental. The whole of mankind depends for life upon this basic operation. It has been so for thousands of years. So important is ploughing that if the ploughmen laid their ploughs aside and failed to plough in the appropriate season then no seeds could be sown and no crops would be grown. The result of this lapse would be no food and the death of mankind from famine before the subsequent harvest of the next ploughing season could be garnered.

Success in any form of endeavour whether it be work or play, depends upon how we begin. This is particularly true in respect of the land. All subsequent cultivations and the economy of doing them depend upon the standard of the initial ploughing. Faults in ploughing can be camouflaged with disc harrows and other tillage implements but their effects are seldom removed even at the expense of extra time and labour. It is, therefore, not only commendable but necessary that a ploughman endeavour to be a perfectionist. There are more good reasons why his furrows should be straight and close, regular and even, than because being so they are pleasing to look upon. It is natural to have a pride in the skill and craftsmanship of good work. This attribute is well pronounced in the good ploughman. Standards of ploughmanship are set by making comparisons with each other's work.

In many countries for many years ploughmen have found delight and friendship in their localities by ploughing together on the same field in their local ploughing matches. At these they select their local champions. By having formed national ploughing match organisations in their respective countries local champions are enabled to plough in provincial or state championships and winners of these in the National Championships.

Through the auspices of the World Ploughing Organisation, an international body for the promotion of the annual World Ploughing Contest in a different country each year, the national champions of all countries are enabled to plough together in order to select from amongst themselves one whom they can call Champion of the World.

The World Ploughing Contest developed out of the desire of ploughmen in many parts of the world to widen their horizons, to exercise their skill with fellow ploughmen on other soils and, also, out of a desire to give hospitality to each other and so share companionship and friendship.

Every year since 1952, when the World Ploughing Organisation was formed, each participating country has sent its two national champion ploughers to compete in the World Contest. These competitors were not chosen by a Board of Selectors. They were not selected by interview or on the recommendation of testimonials. No enquiries were ever made as to who they were or what was their background, or their education, or religion, or politics. They became known entirely by their own endeavours as good ploughmen. They achieved distinction and the honour of representing their respective countries in the World Ploughing Contest because of their good workmanship with the plough. And it is a happy realisation that the plough proves to be a very selective instrument, for those who have participated in the World Ploughing Contests have proved to be good world citizens.

They have ploughed the fields together. They have been honoured and fêted by nations and cities, by governments, by princes and statesmen, by farmers and manufacturers. They have been admired not only for their skill in their craft but, particularly, for their unaffected, natural modesty, their good manners and warm friendliness. They have been appreciated as people to people ambassadors. It is a true saying that good ploughmen make good citizens. They may be of many nationalities but they are citizens of one world. When you meet the national champions, in the happy group that they are you find no feeling of nationalism but a natural realisation that these are, Ploughers of the same earth. These sons of the soil are a brotherhood. Whilst they plough the good land that man might live they are not so lonely as the poets have imagined.

Chapter 1

TEACHING GOOD PLOUGHING

Within the agricultural fraternity and its ancillary occupations there were a few sceptics who saw no value in ploughing matches and viewed the plough as an out-dated and unnecessary tool.

To dispense with the argument I re-produce an article I wrote for the *Journal of Agricultural Engineering Society* of the Department of Agriculture of the New South Wales Government in Australia. This was in response to a challenge by that journal's editor, Graeme Quick.

I had previously written a Guest Editorial in *Soil and Tillage Research Journal* of the International Soil and Tillage Research Organisation (ISTRO) propounding the idea of teaching mouldboard ploughing in developing countries. The ISTRO article was roundly criticised and challenged in the Australian journal.

First of all, here is the text of my editorial as it appeared in *Soil and Tillage Research Journal* (6/1985).

"Teaching Good Ploughing In Agriculturally Developing Countries

From time immemorial the plough has been the fundamental tool of arable farming and continues to be so. Its global function is to unleash the fertility of the earth. By tradition it is a symbol of a world-wide fellowship who use the plough in their peaceful endeavour to till the soil into seedbeds in which planted seeds germinate and grow into crops of food.

The soil is man's most precious asset. When husbanded carefully soil can produce bounteous quantities of food. From well-developed farmlands the amounts are even surplus to the needs of both man and beast. In other areas of the earth there are potentially productive soils which have yet to feel the plough. Through lack of husbandry many soils are unyielding and the dwellers thereon hunger and starve. Nature surely intended that mankind should survive and thrive from the production of his native patch. Unfortunately, in several places man's inhumanity to man has introduced diverse forces which have disrupted Nature's plan.

Transferring stockpiles of surplus food to famine areas cannot maintain a population forever. Such essential and worthy action can only be a rescue operation in times of disaster. Scientists tell us that at any one time there is only seven weeks' supply of food in the world. This implies

1

that every seven weeks mankind depends upon the harvesting of food crops somewhere in the world in order to replenish and maintain an ongoing supply to ensure our survival.

If, and it is an unthinkable chance, the ploughmen of the world laid aside their ploughs thus failing to till the soil in due season, then no seeds would be sown and there would be no crops to harvest. In consequence, the seven week world food stock would all be consumed long before the ploughmen could plough again, and during the waiting period from seed-time to harvest the demise of mankind would be almost, if not entirely, complete. So dependent are we upon our prompt and timely co-operation with the seasons of Nature that to omit one harvest would be fatal.

Tragically, this has become the fate of populations in the famine areas of the world. Emergency transfers of food surpluses will help to rescue some people in the short term, but for the long term survival of the rescued attention must be devoted to the soil under their feet. In this respect the World Ploughing Organisation works to promote a better understanding of the value of soil cultivation practices. Good ploughing promotes better crops which produce more food.

The role of WPO can be described as both educator and inspirer of sporting instinct. Ploughing matches have been introduced and adopted in many countries as a popular and expeditious method of learning by watching, and learning by doing, through the spirit of friendly competition. In agriculturally developing countries, the message of WPO is to explain that all the science and technology, mechanisation and motive power, embodied in the farm tractor and the plough is directed to one vital purpose represented by the "end-product" of the most basic of all arable operations. That "end-product" is the seed-bed.

It is a tradition that country folk enjoy to test their skills one against another and, at the same time, learn from each other's experience and expertise. It is not surprising then that those who learn to plough discover that the soil presents a fascinating challenge. To meet the challenge of the soil they acquire both skill and art. The setting of the plough needs to be as meticulous as the tuning of a violin. The action of the plough must combat weeds, bury trash, introduce air and moisture circulation at root depth and present a maximum surface of soil to the atmosphere to enable sun, wind, rain, frost and snow (depending upon climate) assist in forming the crumble and porosity essential to aid plant root development.

In the final analysis the standards of ploughmanship are determined by the least number of tillage processes which have to be made using harrows etc., to achieve a satisfactory seed-bed and, at harvest, the cleanliness and yield of the crop. Competitive ploughing establishes the standards in the first place. Side-by-side, the ploughed plots can be

2

compared for faults that would obviously be detrimental to growth of crop whilst favourable to growth of weeds. Also, the economy of the ploughman's work can be assessed in a variety of aspects. The lessons are learned by all participants. WPO has encouraged the adoption of ploughing matches in several countries, additional to thirty-two affiliated national ploughing organisations, whose champions compete in the World Ploughing Contest, held annually in a different country. Participation is principally a question of finance to cover the prize which pays the travel costs of the champion and runner-up to the Contest.

WPO has other plans it would like to fulfil. Those plans include sending the World Champion ploughman, or a small group of World Contest competitors, to plough and demonstrate the art and skill of ploughmanship in those desperately hungry areas of the world. To show how their native soils can be made productive. To teach ploughing and inspire the achievement of good standards through the fellowship enjoyed in friendly competition. These plans, and also the staging of the World Ploughing Contest in an agriculturally underdeveloped country, can be achieved with support of sponsors and co-operation of appropriate agencies. A comparatively inexpensive launching exercise could establish a rural-wide activity to achieve the management of the soil to produce food."

The Challenge

"World Ploughing Competitions For The Underdeveloped?

I would like to challenge the World Ploughing Organisation (WPO). The Guest Editorial in *Soil and Tillage Research* (July 1985) contributed by WPO's Director, exposes itself to criticism. WPO's Director Alfred Hall maintained that (quoted in part):

'The role of WPO can be described as both educator and inspirer of sporting instinct. Ploughing matches have been introduced and adopted in many countries as a popular and expeditious method of learning by watching and learning by doing through the spirit of friendly competition. In agriculturally-developing countries, the message of WPO is to explain that all the science and technology, mechanisation and motive power embodied in the farm tractor and the plough is directed to one vital purpose represented by the end-product of the most basic of all arable operations. That "end-product" is the seed-bed . . .

WPO has other plans it would like to fulfil. Those plans include sending the World Champion, or a small group of World Contest competitors, to plough and demonstrate the art and skill of ploughmanship in those desperately hungry areas of the world. To show how their native soils can be made productive. To teach ploughing and inspire the achievement of good standards through fellowship enjoyed in friendly competition. These plans, and also the staging of the World Ploughing Contest in an agriculturally undeveloped country, can be achieved with the support of sponsors and the co-operation of appropriate agencies.'

If there is anything farmers in Australia seem to have learned from their efforts to make a living from this tough environment on the mainland, it is that European cultural practices do not readily translate to our own. It took a few decades to get over the European-style mouldboard plough, once the stump-jump concept took hold, then to this was added the disc plough and tined equipment to make those implements suitable to dryland farming on our shallow soils.

Today even the disc plough is out of favour as we seek to produce crops with minimum erosion tillage in order to preserve topsoil. There is no place here in broadacre dryland agriculture for the mouldboard plough. Furthermore, perfectly clean seedbeds are nothing to boast about. One heavy summer storm can scour clean-tilled bare paddocks beyond recognition.

Ploughing competitions with mouldboards have virtually nothing to contribute to mainland farming (Tasmania is excepted). Based on this conclusion it is submitted that mouldboard ploughing is likely to be a dangerous prescription for underdeveloped arid areas.

Does the ploughing competition have a role in the third world? Agricultural Engineer, Ken James, of VCAH Dookie, has recently returned from a stint in the Sudan. He finds little use for the idea of holding a ploughing match in a place such as the Sudan. Ken James writes:

'Director Hall mentions the end-product of ploughing as the seed-bed, but definitely the end product in these countries must be food. Any input to farming must be measured by some output and I think ploughing must be measured in yield response, not nice straight furrows or seed beds. Yield in Sudan was much more dependent on factors other than ploughing, i.e. rain, weeds, variety, time of sowing, plant population and management.

If help is to be given to these needy people some rules should be followed. I saw many expensive projects in Sudan fail, not for lack of money but for more mundane reasons. Any project should concentrate its resources in a selected area and provide back-up for its people. Educating locals is essential and a realistic time scale selected. A ploughing contest in Sudan would be of little value without follow-up work in planting and growing crops and teaching people overall farm management skills.'

4

The WPO may have noble objectives in mind, but they might as well organise a competition on the moon for all the good it would do in the third world – or on mainland Australia. Is mouldboard ploughing even essential in Europe? How about sponsorship for a competition for best machinery and operators instead?

WPO's response to this open challenge is invited through these columns."

G R Quick (*Journal of Agricultural Engineering Society* (Australia) Vol 14, No 2 1986)."

The Response From WPO

"We are not unaccustomed to your challenge! All down the ages the plough has remained the first tool of arable farming. Probably the oldest plough in existence is the one discovered in 1927, and said to date from 2000 BC, which is in the museum of Lower Saxony in Hannover, Germany.

The pioneer of arable agriculture on Australia's harsh, demanding soil was the plough constructed and used by John Ruse at the time of the first-ever English settlement on the east coast. A replica of the Ruse plough is now Australia's National Ploughing Championship trophy.

Australia's first factory-made plough was imported by the grandfather of the late Hon Thomas W Mitchell, CMG, MA, of Towong Hill. In the early settlement days ploughing matches were a feature of life, particularly in south-east Australia. Around the 1850's, so I understand, ploughing matches attracted more attention than did agricultural shows.

Ploughing matches are still held in Australia culminating in an annual championship final from which the winners proceed to the annual World Ploughing Contest held in a different country each year.

So let us start at the beginning and acknowledge that the plough has been fundamental to Australian agriculture.

The use of disc ploughs and shallow surface tillage tines with the object of leaving trash on top to lessen soil erosion from wind-blow is a feature of the "youngest" of arable areas of the earth. In contrast, over most of the earth's surface (i.e. the "old" arable areas) the soil, is deep enough for digger ploughing. Admittedly, in the most arid areas the tillage was never deep, but it was a plough that scratched the surface soil into a seed-bed even in Biblical times.

From what I have seen of the broadacre dry lands, where a clean seed-bed is not regarded as important, farming the land at all is a high risk business fraught with either drought or a sudden scouring rainfall or a tempestuous wind carrying away soil and seed. They are brave men who farm the land from which the yields per hectare are very low compared with those of the intensively arable lands.

In June 1986 the 33rd World Ploughing Contest was held in Canada's prairie province of Alberta where wind erosion is also a problem and late snowfalls an additional hazard. In this land of "no plough" fifty ploughmen from twenty-five countries ploughed both stubble and grass plots to a high standard and to the acclaim of a vast concourse of prairie farmers.

Ploughing competitions with mouldboard ploughs are features of rural "education and inspiration" in the African countries of Zimbabwe, Kenya and Morocco. The mouldboard plough is a good training implement because its successful operation demands a skill that must be seriously learned.

The quality of soil tillage achieved by mouldboard ploughing in Zimbabwe has inspired a spin-off contributing to development and improved performance of the disc plough in the hard, dry, abrasive soils. Basing their research on the action of the mouldboard, Zimbabwe farmers have developed specially shaped discs fitted with scraper-inverters which cope with depth and side movements to increase or decrease inversion and packing and, when required, completely bury six feet of tropical weed and crop trash.

Certainly, ploughing competitions have played, and do play, an important role in those parts of the third world. In addition to the countries already mentioned competitions have been adopted in Zambia and India, whilst in Thailand there stands a monument commemorating a ploughing match, on which is inscribed "Civilisation follows the Plough".

Referring to the quoted comments of Ken James, it would seem that he has not read the whole editorial which appeared in *Soil and Tillage Research Journal*. If he had he would have read that "good ploughing promotes better crops which produce more food". He would also have avoided quoting the "end-product" out of context. The article specifically refers to the mechanically produced "end-product" of this basic of all arable operations (involving so much science, technology, mechanisation and motive power) as the seed-bed in which "planted seeds germinate and grow into crops of food". (Refer to initial paragraph of ISTRO article).

The object of making a seedbed with whatever tool – be it badza, spade, plough, disc or tine – is to provide the environment in which the seeds can take root and grow into food. Apparently, Mr James does not regard "input" of good ploughing and well-prepared seedbeds as contributing much to the "output" represented by yields? If he does not start with the basics of good food production how does he deal with the other factors mentioned, i.e. rain, weeds, variety, time of sowing, plant population and management? Does he not co-operate with Nature?

6

He apparently does not regard the plough as a management tool? Yet one can find farmyards and scrapyards littered with discarded out-of-date tillage appliances which, in their time, have been pronounced to be the latest, easiest, fastest, and most economical soil management machines destined to make the plough obsolete. And, still the plough persists century after century as the principal basic implement for the management of soil. To the east of Berlin in Germany there is an area traditionally noted for its poverty stricken sandy soil, once thought to be capable only of growing a few conifers, which has been transformed by amelioration ploughing achieved with a two-bodied plough. One body ploughs deep and the other shallow in such a way that the light blow-away topsoil is changed place with the somewhat more robust sub-soil. Such management has made it possible to grow food where none grew before.

For evidence of today's market for ploughs a good example is a plough factory in Norway which in the past couple of years has doubled its production capacity to meet world-wide demand.

Unfortunately, Mr James has written his rather critical comments unaware of other expressions of World Ploughing Organisation policy on the subject of practical aid towards long term relief from hunger. WPO has stated that the most important follow-up aspect will be support from international development agencies to ensure that the operators we train shall have equipment, seeds and fertilisers to carry on with for the future, so that they can become self-supporting in the long term. There must be no failure to take up the initiative in an on-going manner immediately.

Finally, it is an acknowledged fact that ploughing competitions held in developing countries have significantly improved the standards of soil tillage and crop production. They have encouraged an understanding and care for both soil and product and a pride in achievement by the ploughmen.

These competitions are fundamental exercises for those who live and work on the land. They demonstrate that good ploughing is minimum tillage at low cost whether it be done by ox, horse or tractor power.

For your reference to the moon, Mr Editor, if you are to suggest that the moon or a planet could become the Fourth World then might it not be a good idea if the settlers there also have a plough, like John Ruse had at the time of settlement in Australia – or at least a spade?"

(Alfred Hall).

Misguided Criticism

Misgivings about the efficacy of ploughing competitions were, of course, personal opinions expressed by a few detractors who ought to have known better. Mechanisation with increased speed, made real-life horse power redundant and reduced skilled manpower on the farms when, in reality, skilful ploughmanship with new equipment was needed.

I recall two examples. When the British National Ploughing Championships were established it became necessary to compile a glossary of terms common for all to understand, eg. that sod-ploughing in one place is the same as lea-ploughing somewhere else; that a mouldboard is also a breast, and a sole-furrow is also a scolt. There was the question of what to judge and how to judge. The aspects of ploughing include not only the reason for straightness and parallelism. Important, too, are constant depth and width of furrow, angle and uniformity of furrow slices, to mention only a few. There was the question of how to assess the value of each aspect and of the overall value of the finished ploughed plot. Must the assessment be graded "very good, good, medium, poor, bad"? If so, of three equally good looking plots, is one good better than another good? Or, by what measure is one worse than another?

Obviously, assessments have to be scored by arithmetic. A good looking plot may be admired, but it may only look good until the judges walk across it when they will perceive its faults. They will recognise whether or not the furrow slices have sealed off future weed growth and if enough soil surface is presented to the atmosphere to enhance eventual harrowing into a crumbly seedbed.

Whilst guest speaker at a ploughing society dinner in Yorkshire years ago, I was asked to explain the method of judging and scoring adopted by the British Ploughing Association. The system was simple; points up to a maximum of twenty were awarded to each of five aspects (crown, packing, seedbed, uniformity, finish) providing an overall maximum score of one hundred points. The results showed in percentage form how much any plot was better or worse than another. The analysis also indicated to each competitor where his plot was good and where it could have been better.

In his response the Chairman of the Society pooh-poohed the idea of judges awarding *points* to decide the winner and was absolutely opposed to the system. In his opinion anyone who knew anything about ploughing could very quickly choose the winner without having to count figures!

Nowadays, the points award system is meticulously employed for the benefit of all competitors who check from the assessments how their strengths and weaknesses compare.

Another example of misguided criticism came, surprisingly, from an agricultural machinery manufacturer who did not favour ploughing matches. In his opinion if a farmer wanted to know how to plough he could call upon his plough or tractor dealer who would send a demonstrator to show him. But who would be the demonstrator? Probably, a workshop mechanic? Not a ploughman who had qualified on the farm field and proved his skill in competitions.

This particular manufacturer was late to appreciate that farmers take the opportunity to judge the advantages and disadvantages of various makes and models of tractors and ploughs under actual working conditions, side by side, on the same field being operated by ploughmen. At a ploughing match they can observe equipment operated by ploughmen in competition and also assess manufacturers' working demonstrations on the trade plots. Nowadays, wise dealers and manufacturers use experienced ploughmen to demonstrate their products.

These controversies are long past. Misgivings were like small cobbles which get pushed aside in the furrow of ploughing progress. The analytical method of judging is adopted world-wide, and the participation of the agricultural machinery industry is involved not only in exhibiting and demonstrating at ploughing matches but very much in co-operating in the promotion and support of ploughing competitions and ploughmen.

Teething trouble can be expected whilst a young organisation grows. Those against step aside, or eventually become in favour. However, when an organisation succeeds and becomes internationally prestigious, irritation of a different kind is probable from sources without any interest in the purpose or objects of the organisation. These are the political predators who are prone to latch on to an international event, where there are large crowds, and distribute leaflets, display banners, orate and antagonize, creating international sensationalism for their own publicity.

Chapter 2

LAUNCH OF WORLD IDEA

During the war years 1939 to 1945 tractors were as important on the home front as guns were on the war front. An intensive food production programme was imposed on every acre of land whatever its fertile rating. Gardens and even golf links were exploited in the campaign to "Dig for Victory".

Tractors arrived from America to augment and ultimately replace the power of horses on the farms. Horse ploughmen had to learn to drive tractors and men and women who learned how to drive tractors had to learn how to plough. The new-fangled tractor driver had to adopt the skills of the traditional horse ploughman. Consequently, ploughing during the war years was generally of indifferent quality compared with that of the pre-war match-trained horse ploughman.

During the immediate post war years farm horses rapidly reduced in numbers. Tractors and ploughs were mass-produced. No more did the local blacksmith build ploughs to the specification of farmer customers. The time became opportune to devise new competitions for tractor ploughing that would re-establish and maintain good standards of workmanship. Looking further ahead it seemed feasible that tractor ploughing matches need not be limited to local events but might become national, and perhaps international.

This idea led to a meeting convened by the Workington and District Agricultural Society with the Northern Ireland Ploughing Association on 15th February 1951 in the Belfast office of the Ulster Farmers' Union. It was planned expediently to coincide with the visit of a small group of Canadians, members of the Ontario Plowmens Association, who had won a prize visit to the British Isles and Europe, and fitted well with the planned intention agreed between the Workington Society and Sweden's ploughing match committee. The venue was opportune whilst ten Young Farmer Club members from the Netherlands were also in Northern Ireland as guest participants in the Northern Ireland ploughing match. Their leader, Arie Stehouwer, admitted they had spent only one day practising together in Holland under the guidance of an instructor who, like themselves, did not know the rules for ploughing. Only when he discovered that British,

11

Swedish, Norwegian, Canadian, Swiss and Irish farmers were present did he realise there was something more on the agenda than the Northern Ireland ploughing match.

The meeting was an informal get-together with Workington Agricultural Society's Chairman James Lancaster in the Chair, supported by his Vice-Chairman Alf Bowe and Northern Ireland's Vice-Chairman Archie McFarlane and for the Ontario Plowmens Association Vic Porteous. The purpose was to promote the idea and consider the possibility of founding an international body to establish ploughing competitions on a world scale and establish a World Ploughing Organisation with an affiliate in each country. Each affiliate would hold its own national, federal or state championship match from which the winner would go forward into a world championship contest.

The prospect, already communicated to several regions around the world by the Workington Society, had aroused a lot of favourable interest. It would now be helpful to hear if the idea appealed to the ploughmen and their colleagues present at the meeting. Comments and suggestions were welcome for collation and further study. Canadian Vic Porteous said the idea and consideration of rules and standards of work had been a point of mutual discussion between Ontario Plowmens Association and Workington Agricultural Society. It was proper that a single organisation be responsible for deciding rules and for organisation at an international competition. The Canadian desire was to establish a world contest judged by an internationally agreed scale of points. Full support was expressed by John J Bergin on behalf of the National Ploughing Association of Ireland, by Herbert Hughes from Wales, and Henry Kissack from the Isle of Man.

However, Mr Jameson, General Secretary of Ulster Farmers' Union, was not so sure. Northern Ireland, said Mr Jameson, already had a ploughing match with international connections. If the Workington Society was to organise a match on an international basis did this mean there would be two "internationals"? Also, if a British Ploughing Association was to be formed would that imply that the Northern Ireland Ploughing Association with its sixty affiliated local societies would only have equal standing with individual cross-channel societies? So far as rules were concerned the NIPA had held meetings at which every person attending had an opportunity to criticise and offer suggestions which were embodied in NIPA rules when possible. He suggested that England, Scotland, Wales and the Isle of Man should each form their own ploughing Associations like in Northern Ireland.

At this contentious stage in the meeting, what was thought to have been clearly understood by all had to be reiterated. It was outspokenly declared that in no way were the aspirations of the Workington and Ontario organisations expected to cut across the interests of any other organisation. They were intended to enhance the interests of all ploughing match societies and ploughmen. Certainly it was the intention of the Workington Society that Great Britain, of which England, Scotland and Wales are the components, should have a national ploughing association and a national ploughing match of which neither Northern Ireland nor the Republic of Ireland would be parts, since each already had its own established fully representative championship ploughing match.

Apart from a tedious homily upon preserving the status of the Northern Ireland Ploughing Association the main thought-provoking contributions dealt with reaching agreement on a standard style of ploughing and with rules for international competition. Much dissatisfaction was expressed about the disregard of competition rules; especially a condemnation of handling to shape the slices so as to hide the faults of not good ploughing, and a plea to disallow the ploughman being assisted by helpers to handle the plough and titivate the furrow slices.

Whilst no opinions were voted upon, since no provision was made for resolutions at the outset, we felt assured that all present had gathered a good knowledge of views on rules and ploughing styles. What was needed was a detailed description of the type or types of ploughing most suitable for international competition. Also, should a standard type of plough be defined? A decision would have to be made whether or not to ban the use of boats, seamers, handling and assistance given by other persons when ploughing. Would judging be based on "artificial" perfection or on untouched practical work? A list of persons suitably qualified to be accepted as international ploughing judges was needed. Thought had to be given to compiling a scale of points for the assessment of all the aspects of ploughmanship.

In the meantime, the urgent requirement for Workington and District Agricultural Society was to convene an open meeting for founding a British Ploughing Association. Thereafter the work of draughting international rules could begin. To address the world could be done more confidently from a national platform.

Phases of the WORLD-CHAMPION-SHIP - Planning

Commentary By

"Tongue in Cheek"   Leo Parrer (Austria).

1 ENTHUSIASM
OF ALL
PERSONS !
... STILL ENOUGH
TIME...!!!

2 ENTAGLEMANT
? WHO ? HOW ? WHAT ?
WHERE ? WHY ?

3 DISILLUSIONMENT
TOO MUCH WORK !!!

4 SEARCH FOR CULPRITS
WHY HAVEN'T YOU
DONE THAT..?!!

5 PUNISHMENT OF
NON-CULPRITS
" MISTAKE AFTER
MISTAKE "!!

6 HONOUR TO
NON-PARTICIPANTS
YOU HAVE DONE YOUR
WORK   WELL !!!

Chapter 3

INAUGURAL ENTHUSIASM
and POLITICAL OBDURACY

To advance from local to national and international ploughing competitions required a common style and a common formula for judging. Most important, these first principles had to be defined both practically and scientifically in relation to the purpose of good soil husbandry. They had to include all the aspects of good ploughmanship that are a pre-requisite for minimum tillage. Therefore, competition ploughing must reflect a high practical standard to set the criterion for good ploughing on the farm. Ploughing matches are not only sporting, they also educate.

On 17th May, 1951 the Workington and District Agricultural Society convened a meeting in Leeds, Yorkshire, attended by representatives of ploughing match societies from England, Scotland, Wales and the Isle of Man. At this meeting was founded the British Ploughing Association. Six months later the first British National Ploughing Match was held on 14th November 1951. The venue was Newton Kyme near Tadcaster in Yorkshire. Thus the BPA was established.

Plans for the desired World Ploughing Contest could henceforth be promoted with the aid of the new status nationally representative Association. The next act was to invite supporters from other countries to attend an international conference with the object of founding a World Ploughing Organisation.

Where and when to hold a conference was quickly decided. In attendance at the Tadcaster event was the Mayor of Workington, Alderman Mark Nilsson MBE who asked where was it intended to hold the conference? Upon being told that no thought had yet been given to choice of venue, the Mayor suggested Workington and offered the facilities of the Town Council Chamber, a Civic Reception and Dinner for the international delegates.

So it was that the inaugural conference of the World Ploughing Organisation was held in Workington, Cumberland on 5th February, 1952. A Provisional Governing Board was agreed of one member each from Great Britain, Canada, Sweden, Netherlands, Republic of Ireland and Finland. Two further meetings were held respectively at Stirling and Bridge of Allen, in Scotland in November of the same year, to coincide with the 2nd British Ploughing Championships, when the membership was increased to include one member each from Norway and West Germany. The founding and Constitution were ratified and the Board agreed to hold the first World Ploughing Contest in Canada at the invitation of the Ontario Plowmens Association. The dates for this great new event were from 6th to 9th October 1953 when the United States of America and Denmark also became represented on the WPO Board.

In that same year 1953 on 1st June was introduced the first Federal German Ploughing Match at Holweide, near Cologne. The motivator for the adoption of a national ploughing match in Germany was soil scientist Walter Feuerlein through whose leadership were established one hundred local competitions and thirty-six selection matches at regional level within six months of the founding of WPO.

At the invitation of Walter's committee John Bergin and I were observers at the Federal Championships. John was popularly referred to as "J J" among the ploughing fraternity. He was founder of the National Ploughing Association of Ireland and valued the ploughman's caring for the soil which he warmly expressed in his delightful poem "The Song of the Plough". Unfortunately, J J's understanding and dedication to the symbolic, peaceful plough was pervaded with political dogma. He was argumentative. Nevertheless, he promised to donate a trophy to WPO to be presented to the Ministry of Agriculture in the country from which came the world Champion ploughman.

At the time of our visit, Cologne and district was in ruins from wartime air-raids. Most shops were limited to using temporarily restored ground floor rooms. Upper stories were rubble. Only the imposing Cathedral looked majestically strong. We were guests at separate farms out in the countryside. It was already night when Professor Helmut Frese, of the Volkenrode Soil Research Station delivered me to my hosts. The house was at the side of a wide courtyard encompassed by barns, byres and stables. After a warm welcome and very satisfying meal, the farmer suggested I accompany him and his wife to visit friends for the remainder of the evening. Upon reaching for my coat he said it was not needed. I expected we would go by car. Instead we walked a few yards along a passage into another room. Here, in the same house, lived their friends. They were a bombed out family in temporary residence. So, too, were several other families billeted in different parts of the house and the farm buildings. All were

16

awaiting new homes to be built. After a pleasant couple of hours conversation, cakes and a nice wine it was time for bed.

Next morning, I rode with Helmut to a country inn – gasthof – where Walter Feuerlein and his ploughing committee were already waiting with J J. As we stepped out of the car Walter greeted with the words "We have special news for you. Hilary has climbed Mount Everest! That is a present for your Queen on her Coronation Day!" Walter and his German committee then sang "God Save the Queen". The date was 3rd June, 1953.

After attending the deep-digging ploughing match, (exclusive at the time to short, concave mouldboards), having admired the shirt-sleeved Minister of Agriculture plough with horses a plot of "Opening Ceremony" furrows, and congratulated the first Federal Champion Ploughmen Horst Jahn and Roman Dondorff, J J and I had opportunity to visit the large German Agricultural Show (Deutsche Landwirtschaft Gesellschaft), an event of similar prestige to the Royal Agricultural Show in England.

Sad to relate, J J and I had different interests. I understood he was both a farmer and an engineer and expected we would have a common interest in what we saw as we wandered around the great variety of livestock and farm machinery exhibits. But J J expressed no particular interest and irritatingly entered upon the subject of Irish politics! He said that whilst the six northern counties of Ireland had a ploughing match conducted by the Northern Ireland Ploughing Association he wanted me to know that the National Ploughing Association of Ireland, of which he was the founder and managing director, represented the whole of Ireland and was the only ploughing association in Ireland entitled to enter Irish ploughmen in a World Ploughing Contest.

I could see no good purpose in political argument. To me a ploughman was a ploughman wherever in the world he lived and ploughed. Surely, in their peaceful profession ploughmen shared fellow feeling in their hearts and would desire to plough their best in a prestigious world event.

All I could say was " Well, J J, that's a matter for your NPAI and the NIPA to agree upon. As a matter of fact I'm not sure that areas of ploughing associations affiliated to WPO need necessarily be defined within political boundaries. After all, we cannot define ploughing areas of the world as coinciding with political frontiers. If we look at a soil map we see some homogenous arable areas overstretch political boundaries whilst others lie well within them. There are areas within large countries that might be considered separate affiliate areas. The real frontiers of the plough are water, rock and frozen wastelands. Personally, I would like to give every ploughman everywhere a chance to participate with the brotherhood".

J J was not in tune with my theme. In his opinion the Northern Ireland Ploughing Association should not be recognised and ploughmen representing the six northern counties should not be eligible to compete in a World Ploughing Contest. I pleaded, "J J, that is not for me, personally, to decide. We are only at the stage of finding where ploughing matches are in the world, trying to form them where there are none, and to collaborate with ones already established. Now, J J, let us look at the horses and ploughs and tractors. If we stand here we'll never see the Show".

Mr Bergin would not be put off. He discoursed upon terminology. Although the Irish Post Office franked letters with the name "Eire" he was at pains to point out that neither Eire nor Southern Ireland were correct. Members of WPO must be told to use the name "Republic of Ireland". I was becoming bewildered for I never found happiness

One-man Thresher, among multi-exhibits at DLG show, Germany

18

associated with politics. We began to stroll towards the livestock exhibits when J J suddenly lifted my camera from off my shoulder and said, " I am in possession of a bit of your property. What do you do about it?"

"Well, J J, I am making a new arrangement with you. Like a good friend you will now restore the camera to my shoulder. And, so that we both shall see what we each want to see, I will go this way around the show and you will go in the opposite direction. We'll meet again together with our German friends, as arranged, at three o'clock in the International Pavilion". And that is what we did. Germany was no longer making armaments, metal was used to build farm machinery and there were two-hundred and thirty-two types of tractors and at least two thousand models of ploughs in Germany, and a lot of them were to be seen at the Show.

Chapter 4

MID ATLANTIC INTRIGUE

On 20th September 1953 two champion ploughmen from each of seven countries and one each from Finland and Northern Ireland, together with their respective WPO Board Members, sailed from Liverpool on the Canadian Pacific liner "RMS Empress of Australia". Their ultimate destination was Coburg, Ontario to participate in the first World Ploughing Contest. The total number of competitors was twenty-two including two Canadians and two Americans who, of course, were already on the other side of the Atlantic Ocean.

Their names were:
Leslie Dixon and Reg Hogg, Great Britain.
Jim Eccles and Bob Timbers, Canada.
Odd Braut and Olof Nedburg, Norway.
Alan Helin and Per-Gote Jonsson, Sweden.
Wim de Lint and F T L van Gaalen, Netherlands.
Tommy McDonald and Ronnie Sheane, Republic of Ireland.
Egil Anderson and Erik Bie, Denmark.
Horst Jahn and Roman Dondorff, Federal Germany.
Graeme Stewart and R C Cummins, USA.
Hugo Sintonen, Finland.
Bob Carse, Northern Ireland.

Bob Carse was the only one who was unaccompanied by a representative of his ploughing association. Although the Northern Ireland Ploughing Association had not yet decided to affiliate with WPO, Bob by virtue of being current Northern Ireland Champion received his travel award direct from Esso Standard and Imperial Oil Company sponsors. Without the voyage being sponsored by Esso and the generous Canadian hospitality the World Ploughing Contest would have been a non-starter. International currency restrictions forbade taking more than five pounds sterling out of the country. European nationals required visas with their passports. German competitor Roman Dondorff was a problem because he was bred and born in East Germany but managed to escape across the Iron Curtain from the Communist regime into the west and found work on a farm. Roman competed in ploughing matches and won the first German championship at Holweide. Not withstanding the fact that he had become a citizen of West Germany the Canadian authorities at first refused him permission to enter Canada. This personal crisis involved much correspondence, pleading and arguing. At the last minute before sailing the Canadian authorities relented with the acceptance of our assurance that poor Roman would continually be in the responsible company and care of WPO and our Canadian hosts who were a very impressive list.

On passage through the Irish Sea the wind blew brisk and salty. Waves were high and troughs between them deep. The world ploughing group gathered on the boat deck for photographs, explored the ship, learned about the social programme for what became a voyage lasting nine days. Deck games were played on top and "horse racing" below. Everyone became acquainted with a feeling of fellowship. The exuberant Roman Dondorff began to climb the rigging and descended sharply when caught by the disapproving eye of an

World Ploughmen aboard RMS Empress of Australia

22

Officer on the bridge. He had to be told that much as we understood his delight to be free there was a limit to his freedom.

Meanwhile the ship headed through the restless North Channel between Scotland and Ireland to plunge into a turbulent Atlantic Ocean whipped by a frenzied Force Ten gale. Next morning, after a good breakfast brave ploughmen strode unsteadily around the heaving deck; typical of a countryman's wont to breathe open air! The crew was bolting down tables and chairs and fixing long lengths of sturdy ropes for hand hold along all the decks and throughout the dining rooms and lounges.

It was soon obvious that not everyone was a picture of robust health. The stumbling deck parade petered out as the great mass of gale-riven ocean tossed our now "tiny" ship, which had looked so huge alongside the Liverpool landing stage, into a shuddering motion of pitching and rolling.

On the third morning the breakfast room was empty and silent except for the crash of waves and the reassuring subdued throb of the engines. Most of the passengers were more than subdued and wishing the ship would stand still if only for ten seconds!

After several days the storm eased. Ploughmen emerged on deck and socialising resumed. In mid-Atlantic there was time to reflect on the pioneer achievement of the past seven years. We were progressing towards our goal. For the first time in history ploughmen from thirteen countries were coming together to plough and compete on land still spoken of as the new world. This was a satisfying thought but the past seven years had taught us that anything worth doing is seldom, if ever, easy. Whilst the ship made its unsteady passage westward an uneasy situation was revealed. Perhaps the first "real-politik" boulder in the furrow since Joel recorded in his *Book of the Old Testament* (chapter 3, verse 10) about the advancement of national interest by assembling in the Valley of Jehosophat with ploughshares beaten into swords.

Our two Irish ploughmen from the Republic asked me to have a private chat with them where we would be neither seen nor interrupted. We agreed to rendezvous in the steerage lounge. No more was said and my curiosity was aroused.

Tommy McDonald and Ronnie Sheane were waiting at the appointed time.

"Why all this secrecy?"
"'Tis J J", they said, "he wants us to pull out."
"Pull out! In the middle of the Atlantic?"

23

"He says we must boycott the Contest, refuse to plough and stand on the headland in protest against the entry of a ploughman from Northern Ireland."

"Does Bob Carse know?"

"Nobody knows. We have nothing against Bob. He is another ploughman like ourselves. We've had so many lectures from the Old Man that we are just miserable."

"What do you want to do?"

"We want to plough."

"And if you didn't plough.....?"

"We couldn't face the folk back home after all the money that's been spent on us and ploughs and tractors provided. We'd have let them down. They'd be mad."

"What do you want me to do?"

"We want you to know our problem and we want your advice."

"Well now, what you do is for you to decide. All I can say is what you already know. You have won the Irish Ploughing Championships, qualified for and been properly entered to compete in the first World Ploughing Contest on everybody's understanding that you will do so. You have been awarded a free trip to Canada where for two weeks you will be guests of Canadian hospitality. Like all the other ploughmen you have embarked upon the attendant obligations in good faith. In a matter of conscience you are free to do what you think. No one will force you to plough. If you plough no one will know you've had a problem. You're on your way, it's too late to get off the ship to go home, that chance was at Liverpool".

"If J J tells you he has withdrawn our entry and that we are not going to plough will you still allow us to plough?"

"Certainly. That's why you are on your way to Canada."

The secret meeting was over. We returned casually to our quarters. The matter was never discussed further. Ronnie and Tommy ploughed as normal along with the other ploughmen. As quoted in the *First Book of Corinthians*, chapter 9, verse 10,

"He that plougheth should plough in hope".

Chapter 5

THE FIRST WORLD
PLOUGHING CONTEST

On 1st October 1953 the "Empress of Australia" berthed at Quebec harbour three days late because of the storm. The voyage through the Gulf of St Lawrence was in complete contrast from the rough ocean crossing. The water was now calm as a millpond. A school of whales lay close to the surface blowing waterspouts. As the sun was setting beyond a shimmering mist the distant landscape appeared in mirage. Early next morning, whilst slowly steaming up the St Lawrence river the sun rose behind us in a blood red sky highlighting the sharp brilliance of the browns, reds, sepias and greens of the autumn foliage. The maple and sumac leaves appeared as though splashed with deep red blotches rained from a crimson sky.

Because of the delay we disembarked instead of continuing to Montreal. Here we were welcomed by the ebullient Mayor Camillien Houde, the Imperial Oil Company for lunch, and to dinner by the Canadian Pacific Railway Company whose business included the ship which brought us, farms, hotels and the Canadian Pacific Airline. Both functions called forth speeches of welcome and novel responses in our six languages with translations.

An interesting revelation came from seventy-two year old Norwegian farmer Arne Braut who was accompanying his forty-one year old competing ploughman son Odd. The Brauts of Norway are a ploughing family, like the "ploughing Dixons" of England, and the "ploughing Wrights" and "ploughing Barrs" of Northern Ireland, and many other families of similar note. For Arne Braut the first World Ploughing Contest motivated the accomplishment of a long felt desire. His brother Iver left the family farm when he was sixteen years old and emigrated to America. He arrived there with "nothing but the clothes on his back and a strong pair of hands".

In Norway Arne had a farm of about forty acres and son Odd had one of similar size. After the Ploughing Contest they visited Iver on his twenty-five thousand acre wheat farm in North Dakota. There the two brothers were together again for the first time in fifty-seven years.

From Quebec our travelling was by Johnnie Stewart's Bus Line

Parade to the Plots, First World Contest.

under the banner "World Ploughmen on Tour". Pre- and post-Contest journeys covered two thousand miles. First to Montreal and onwards to the township of Cobourg where the Opening Ceremony for the first World Ploughing Contest was performed by Canada's Governor-General Vincent Massey. Flags of the nations were flown, bands played and the ploughmen drove in an impressive parade to their respective contest plots. The four-day event of first Ontario and then World Championships celebrated forty years of the founding of the Ontario Plowmens Association. Six hundred Canadians, horse and tractor men, competed during the first two days and the World Contest competitors ploughed stubble on the third day and grassland on the final day.

Thirty-four year old Canadian Jim Eccles won the "Golden Plough" to become the first world champion ploughman with 154.66 points and forty-one year old Norwegian Odd Braut was runner-up with a score of 151.32 points. The lady reporter for the *Cobourg Centinal Star* wrote "the tractor driven by Roman Dondorff from West Germany drew considerable interest because it was an air-cooled Diesel model – the only one of its type used in the match". Diesel-fuelled tractors were virtually unknown in North America at that time. The same writer continued "perhaps the ploughman who captured our imagination most was the one from Northern Ireland. All of our Irish blood thrilled to the large shamrock he wore on his back, and his dark handsome features beamed an irresistible smile when the Irishmen in the crowd gave him a rousing hand-clap".

Our exuberant Roman Dondorff, the refugee from the east, must have been over excited during the pre-contest practice period. He overturned his air-cooled Diesel tractor. This incident posed a problem for Walter Feuerlein, coach and mentor to the two German ploughmen. There was no agent or dealer service facility for this make of tractor in Canada. Unfortunately, it needed a vital repair. Walter, being an enterprising chap, contacted the German Consulate. By good fortune two German immigrant mechanics, acquainted with this make, were found and willingly put the machine into good order.

Walter wished to ensure that the accident be given no publicity and that I have a fatherly but serious word with Dondorff in the belief that he would be subdued by admonishment from a foreign secretary more so than from a German coach. Consequently, Roman was asked to see the WPO Secretary at headquarters. He was told in a very friendly manner how unfortunate was his accident and how fortunate was the repair job. Then he was firmly told to bear in mind that being a

27

Jim Eccles, first World Champion finishing his plot (Imperial Oil Limited)

champion ploughman of Germany he was regarded by his admirers to be an expert. His skill, craftsmanship, care and safe handling of his tractor was expected to set an example to all others.

Roman, relieved and grateful, thenceforth handled his tractor with care, and during the voyage home did not climb the rigging! The last I heard after his return to West Germany was that he became a good farmer and community worker.

More than 100,000 spectators witnessed what became aptly known as the first "Ploughing Olympics". In addition to competition ploughing there was a huge display of farm machinery both static and working on the land. The avenues of trade stands totalled three miles. One particular exhibit was symbolic of "swords beaten into ploughshares". An enormous man-high forestry plough, the first of its kind, was built by Mr Skaerpe, a Norwegian inventor, from steel plate recovered from the German battle ship "Tirpitz", raised from a Norwegian fjord after being sunk in battle during World War II.

Following three days of setting ploughs and practising, two days of competing, culminating with a "Golden Plough" Presentation Banquet for twelve hundred diners, the next seventeen days were spent on a

28

whirlwind bus tour within the Province of Ontario. October is a festival season when the traditional fare is turkey and pumpkin pie; a menu to which we became well accustomed, often twice daily in different locations on our travel. And, every meal was accompanied by numerous speeches all listened to with courtesy but not understood by non-English speaking guests. Sundays were for rest and opportunity to attend a local Church. On one occasion, at Rice Lake, for unknown reason, the Minister failed to arrive but the local congregation and world ploughmen joined in prayer and hymns without him.

Through forest and across plain Johnnie Stewart's bus raised clouds of dust as we sped along dry, unsealed roads. We came to the deep, slow flowing Grand River where the homeland Reserve of the Six Nations Iroquios North American Indians extends six miles wide from each river bank from source to mouth in Lake Erie. The cross-river ferry was an ingenious log-raft winch-hauled on a rope secured to each bank. We were guests of the Six Nation township of Oshweken, social and political centre of the Reserve. The tribal Chiefs of the League of the Iroquios have held their Pow-Wows in their Council Chamber since the year 1874, representing the Mohawks, Senecas, Oneidas, Tuscaroras, Onondagas and Cayugas. Today's Council Chamber is no longer the original Loghouse of 1874 when the illustrious Thyendanagea, also known as Captain Joseph Brandt, led his fellow tribesmen from the USA to settle along the Grand river banks and pledge loyalty to the British Crown. Brandt was a Mohawk and established the Mohawk village of Brandtford.

The Mohawks and their fellow tribes were granted the tract of land as their Reserve for them and their descendants to enjoy for farming, fishing, hunting and forever, without taxes, being the true natives of the continent.

Within one year of settling they built St Paul's Chapel of the Mohawks, also known as the Chapel Royal, and as Her Majesty's Chapel of the Mohawks since for them it replaced their original Queen Anne's Royal Chapel in Mohawk Valley, USA. This was the first religious edifice to be erected in Ontario. Schools were established and also the Mohawk Institute where Indian boys and girls are educated in farming and domestic science.

Community spirit is reflected in an energetic Womens' Institute, for it was at Stoney Creek where Mrs Adelaide Hunter Hoodless founded the first Womens' Institute on 19th February, 1897. A memorial cairn commemorates her birthplace.

Brass bandsmen wearing buckskin and feathers led us in parade

29

Some Six Nations Citizens of the Iroquois

through the village and into tribal dances with beating of tom-toms and chanting. The charming, bright-eyed, ruddy faced, black-haired WI squaws provided a generous and delicious banquet for our "Eleven Nations" ploughmen. Turkey with cranberry sauce, pumpkin and apple pie – several helpings! The Chiefs and Princesses were clad in traditional regalia.

Although a letter from the Ontario Ministry of Agriculture had given advance notice I was surprised by what awaited me personally. After the feast in the Long House, Chief Howard Skye of the Cayugas and Chief Little Bear of the Mohawks and his wife "Princess Watching beside the Village" together with Chiefs Deskeheh, Huron Miller, Red Jacket and Red Cloud mounted the stage. Chief Skye announced that the Six Nations Council had pleasure to invite the General Secretary of the World Ploughing Organisation to become an honorary member of the Iroquios and adopt an Indian name. Did anyone object? No one did. Little Bear led me to the stage and Chief Skye declared "We give him the name Rahgahratwas which means 'the Ploughman'!" Whereupon the 81-year old Princess Little Bear ("Watching beside the Village") placed a magnificent bonnet of eagle feathers, ermine,

The Author is given the name Rahgahratwas

buckskin and beadwork on my head. With a warm handshake Chief Skye said "We welcome you as a brother Mohawk" and gave me the historical register to sign.

Then taken by the hand of Little Bear I was led into a ritual dance with the tribal Chiefs involving a low key chant interspersed with cheery whoops. From nervous observation I rather clumsily managed to stamp and shuffle in unison with my brothers to complete the ceremonial initiation.

Our long and memorable tour concluded with a farewell banquet hosted by the Canadian Government in Ottawa. It was a comfortable, informal affair. Seated with our hosts at large round tables in groups of ten per table we were addressed by Prime Minister Lester Pearson: "Listen fellows! You are all very welcome. You have heard a lot of speeches, but we are having no speeches tonight. Sitting with you at each table is a Cabinet Minister of the Canadian Government I will introduce them to you . . .". He called each Minister in turn by Christian name and the name of his department . . . Agriculture, Health, Foreign Affairs and so on . . . and concluded "They will tell you anything you wish to know; chat with your hosts and enjoy your meal".

The homeward voyage from Montreal to Liverpool aboard another Canadian Pacific sister ship "RMS Empress of France" was seven days of friendly companionship and idle relaxation.

Chapter 6

FESTIVAL IN KILARNEY, POLITICS IN CORK

From an early stage in the series of Contests spasms of political paralysis interrupted otherwise peaceful proceedings. It was like trying to control the intrusion of a creeping nettle in the garden.

The first serious experience was an aftermath of the second World Contest held in the beautiful landscape of Kilarney in Republic Ireland in year 1954. Kilarney was in festive mood, gaily decorated and providing a week's programme of indoor and outdoor entertainment. Marching bands and dancers paraded in welcome to the ploughmen who came from thirteen countries, now including from Belgium and France. Crowds of spectators admired the ploughing but many had little respect for it since in their eagerness to inspect they trampled it underfoot until, after an appeal to genial Police Chief Bart Hart for crowd control, he called out the Irish army who roped off the plots to keep onlookers on the headlands.

World Ploughing at Kilarney, 1954 (Aerophotos)

33

Northern Irishman Hugh Barr became the new World Champion with Leslie Dixon, from England, the runner-up. At the "Golden Plough" Presentation Banquet warm tribute was paid to Hugh Barr by Ministry of Agriculture Parliamentary Secretary Oliver Flanagan concluding with the words " We can plough the land together – north and south – we can live in the land together. My heartiest congratulations and good wishes to the World Ploughing Champion from across the Border".

President of the National Ploughing Association of Ireland John J Bergin said, "I am delighted with the result. Hugh Barr is a very worthy winner and we all wish him very good luck. I think myself it could not have been a better result".

Next morning the town band foregathered to play the international ploughmen on their way from Kilarney. The new World Champion was missing. He had slipped away quietly during the night in his brother's motor car to the farm in the north where his father lay seriously ill.

Hugh Barr, winner in action (Esso)

During the post-Contest sightseeing tour of the south of Ireland, including many receptions, time had to be found for meetings of the WPO Governing Board. The first was held at Cork on llth October 1954 when John J Bergin presented a statement which read,

"We, the Council of the National Ploughing Association of Ireland, acting as a constituent member of the World Ploughing Organisation desire to enter a protest against the acceptance of the entry of a plough team from the partitioned portion of Ireland as the acceptance of such an entry denotes approval by the World Ploughing Organisation of the political dismemberment of Ireland".

Mr Bergin told the Board that to permit Northern Ireland ploughmen to all WPO functions raised a purely domestic problem which ought to be referred to the National Ploughing Association of Ireland.

The WPO Board listened in quiet astonishment, made no comment other than to agree unanimously to take no action. They proceeded to consider applications for affiliation from Pakistan Power Farmers' Association, Circle National des Jeunes Agriculteurs France, FN des JAP de Belgique, the Northern Ireland Ploughing Association and an unknown ploughing society in Algeria. It was shortly after midnight when the meeting ended.

In the morning at precisely 07.20am a letter was slid under the bedroom door of the WPO General Secretary. The letter was read to the WPO Board at their next meeting at four o'clock in the afternoon of Thursday 4th October, 1954 in Jury's Hotel, Dublin. Travelling from Cork to Dublin the ploughmen enjoyed several interesting visits ending with a reception at the Mansion house hosted by cheerful Lord Mayor Alfred Byrne. The Board meeting had to be slotted between a wash and brush-up at the hotel and dinner as the guests of Bord Failté, in company with Minister of Agriculture James Dillon and Minister of External Affairs Liam Cosgrave, at the Dolphin Hotel.

The early morning letter was over the signature of J J Bergin and complained that the previous night's Board meeting was "highly irregular" and "any business done was of no effect". Furthermore, "the National Ploughing Association of Ireland would now withdraw from membership of WPO", in consequence whereof Mr Bergin was now leaving the tour party but would explain his action at the Bord Failté evening function in the Dolphin Hotel.

The Governing Board received the letter with regret, without discussion and turned attention to another contentious conundrum posed by Monsieur Fenaille, from France, who advised that an application from an Algerian ploughing society could not be considered because Algerian ploughmen should first qualify in the French National Ploughing Championships in order to compete in a world contest. Since the Algerian application was couched more in terms of a general enquiry the Board avoided a further political complication by agreeing to "lay the matter on the table".

At this point the proceedings were interrupted by the arrival of a member of the National Ploughing Association of Ireland who announced that the Council of NPAI had met during the afternoon and deplored and deeply regretted the action of Mr Bergin in withdrawing from WPO. Subject to the approval of WPO the NPAI Council had nominated their Mr Coulton to attend the Board meeting as an observer. WPO readily assented to this nomination.

Ever present on visits, in the vicinity of meetings, at receptions and social functions were press reporters thirsting for a sensational item of news. According to newspaper reports international ploughing was embroiled in Irish politics. The "peaceful plough" had become a troubled "painful plough". J J's original promise of a trophy now came

Irish Republican Prime Minister J. A. Costelloe congratulates Hugh Barr, World Champion from Northern Ireland (Esso)

36

into the picture. The massively ornate artwork in Conemara marble and silver, originally presented to the NPAI in year 1938 by Minch Norton and Co Ltd, was to be presented to the Minister of Agriculture in the home-country of the winner of the World Ploughing Championship. The World Champion was Hugh Barr, a citizen of Northern Ireland. Mr Bergin decided to put the trophy into storage.

The sequence of political irritation caused embarrassment for both the Minister of Agriculture and the Minister of External Affairs who regretted that because of it they were obliged to withdraw from attending the international banquet in honour of the ploughmen. Oliver Flanagan, Ministry of Agriculture Parliamentary Secretary was appointed to deputise on their behalf. Champion Hugh Barr returned from the North having visited his sick father.

Whilst dressing in readiness for the banquet there was a knock on my bedroom door. It was Oliver with a personal message from Agriculture Minister James Dillon deeply regretting the turn of events and all that caused it. The Department and the Government were annoyed and apologised for the hassle which was not of their creation and which they deplored. In token of their sincerity and appreciation of our handling of the situation Oliver presented me with an ornamental model of the Irish House of Parliament. With a warm shake of hands we both hastened our preparations to be in time for dinner at the Dolphin Hotel. J J was at dinner, too, but made no statement.

The newspapers reported that J J had submitted a five-point memorandum to the Northern Ireland Ploughing Association as follows:

1. That the organisations on each side of the Border conduct their own championships as before.
2. That a final test be held; a number of competitors to be agreed upon representing the areas.
3. That a one-day annual lea-and-stubble competition be held at alternative north and south venues.
4. That a panel of judges be appointed by a joint committee of both organisations.
5. That the two competitors obtaining the highest points be selected to represent Ireland.

In a letter dated 14 September, 1954, the Secretary of Northern Ireland Ploughing Association, J B Somerville, replied stating that the terms "seemed to us to be very fair and reasonable ...." but ".... I am afraid, however, we do not agree either as to the necessity or desirability at this stage for such amalgamation".

Despite all the time-wasting harassment the Emerald Isle's traditional "hundred thousand welcomes" and hospitality was unstinting and memorable.

Five years later the Republic's ploughmen decided they should no longer be kept from participating in the World Ploughing Contests. John J Bergin to whom is credited the formation of the NPAI in year 1932 had died in year 1958. Miss Anna May Brennan (Mrs McHugh) succeeded as secretary with Sean O'Farrell as Managing Director. With amiable forethought NPAI re-affiliated with WPO and made a goodwill payment in restitution of fees which would have been paid to WPO had membership not been withdrawn in 1954.

Meanwhile, the Northern Ireland Ploughing Association became more outward looking by ratifying affiliation with WPO and nominating Tom Reid to be the first NIPA member of WPO Governing Board.

Ever since the adoption of this happier state of agreement ploughmen from both sides of the partitioning border have participated in each other's ploughing matches as competitors and judges. They share arrangements to transport their ploughs and tractors on the same trucks to travel together when shipped to World Contests held on the European Continent. An example of how those who plough build fellowship bridges across political barriers.

Chapter 7

SHILLINGFORD – ROYAL
and INFORMAL

Thankfully, there were no political boulders to clog the clearly defined furrows of administration at the third Contest in Sweden in the autumn of 1955. Northern Ireland's Hugh Barr won again with Canada's Ivan McLaughlin runner-up. Twenty-four ploughmen competed from twelve countries. Among fifteen thousand spectators were observers from east (Soviet zone) Germany, Yugoslavia and Egypt. The land was the flat fertile holm alongside the Fyris River and overlooked by Uppsala's imposing Cathedral and Castle. On the initiative of WPO Chairman Walter Feuerlein and the organisation of Professor Torstensson WPO's first Soil Tillage Conference was held at Uppsala University to coincide with the Contest. Subsequently, WPO held seven more such Conferences in different countries which led to the participating scientists forming the International Soil and Tillage Research Organisation (ISTRO), now with a distinguished membership in more than eighty countries, a Symposium at normally two-year intervals, and affectionately remembered as a "daughter" of WPO.

The East German observers intimated their intention to attend the next year's Contest at Shillingford in Oxfordshire, England. Their party would include two ploughmen. In the event, however, there was no confirmation of their intention during the year 1956 until the receipt of a telegram at the General Secretary's office on the Contest site at Shillingford announcing that a small group of East Germans (presumably four) on their way to England to attend the World Ploughing Contest were delayed at the Border Checkpoint in Berlin and could we help them to proceed. The next few hours were spent contacting the Foreign and Home Offices in London to get information and to try to solve whatever was the problem. The feed-back was confusing because, on the one hand, the East German reply indicated that they were stopped by the British Control whilst, on the other hand, we understood that the British Military Commission had allowed them clearance for the purpose of travelling to the World Ploughing Contest. We arranged that someone would meet them at London Airport. It was already the first day of the four-day event (National Match and World Contest). They did not come! The assumption was that they were

39

detained by the Communist Border Control whose "iron curtain" it was to keep people from coming west. Perhaps it was concocted propaganda to accuse the "undemocratic" British of not allowing the east to go west!?

There were no other political problems with the fourth Contest, but there was a lot of unnecessary fuss about protocol when the Committee decided to invite a member of the Royal family to unveil the commemorative Cairn of Peace. HRH Prince Philip, Duke of Edinburgh was the Patron of the British Ploughing Association and contributed an encouraging foreword in the Contest handbook but, due to prior engagements, could not attend the event. There was much uninformed debate about how to convey the invitation to another member of the Royal family. Should the request be made through the offices of a county Lord Lieutenant, a Deputy Lieutenant, a Member of Parliament, or direct to the Lord Chamberlain's office? And, who should do it? Perhaps it was a job for a small sub-committee? Who were the people with the right connections? So it went on. No thought was given to the possibility that some member of the Royal Family might dearly love to be invited to the National and World Ploughing Matches! When so many want to be in on the act it is better to let the Secretary do the right thing and write to Buckingham Palace.

Thus it came to pass that HRH Prince Henry, Duke of Gloucester, would be happy to attend the event and perform whatever duties were required. The Committee now determined that the Royal visit must be programmed to a strict timetable. It had to be decided which people should be presented to the Duke upon his arrival and which other VIPs ought to be invited. There were plenty to choose from Mayors, Bishops, Government people, Civil Servants and, not to be forgotten the volunteer workers for the British and World Ploughing matches. The ticklish problem was to avoid getting too many non-functional subjects in the line-out.

Whilst the Committee worried about protocol the fact was overlooked that a Royal person does not necessarily have to be appealed to, or persuaded, to attend an event – even a ploughing match. A couple of days before the Contest the General Secretary's office on the field received a call from the Duchess of Gloucester's secretary asking very kindly if we would mind, and only when convenient, if the Duchess accompanied her husband to see the ploughing! My happy response confirmed it was most desirable and very remiss of us not to have addressed our invitation to both, for which I humbly apologised. The lady secretary assured me the Duchess would be delighted for she had expressed a keen desire to see the Ploughing Contest. This and other experiences have shown me how congenial are members of the Royal family when they have a chance to break through the tight ring of protocol. Yes! After all, there was a member of the Royal family who was wishing to come to the ploughing match all the time!

The President of the British Ploughing Association was the Duke of Marlborough and since the Shillingford event was both a Royal and an International occasion it was necessary to discuss formalities with him and his Duchess. Over tea and cakes at Blenheim Palace Her Grace gave a forthright appraisal of the situation. First, it was important not to forget that the most important people at the four-day event were the ploughmen for without them there would be no ploughing and no purpose. That said we concentrated on how to manage the VI(non-paying)Ps who were to be individually introduced to the Duke of Gloucester and entertained to lunch. The order of presentation and seating required some thought.

The Duchess read the list and mused "You have a problem". When Royalty are present some guests get a bit fussy about where they think they ought to be seated. This is where the Duchess's knowledge of protocol and social precedence came in useful. I understood one of her duties was that of Mistress of the Queen's Robes. And so we prepared a seating plan to give all guests a chance to talk and learn something about ploughing without too much regard for precedence.

When the official caterer was told he was to serve Royal guests he panicked about the already agreed menu. What would the Duke want to drink? Was the Duchess on a fish diet? He had to be told, for goodness sake, to regard everyone in the party coming in from the fresh country air and the smell of newly turned earth as having a healthy appetite and similar taste buds. To the caterer's surprise, and no doubt relief, the Duke required no drink other than after lunch coffee, likewise the Duchess and everybody enjoyed the menu.

The seating plan worked well. Pleasant conversation buzzed around the table and the principle subject was ploughing. My whispered wish to the Duke's Equerry that lunch would not drag on until the ploughmen had almost finished their plots was not missed by the sensitive ears and sharp eyes of the Duchess of Gloucester. During a lull in the conversation she charmingly remarked "I'm sure there is much for us to see and Mr Hall would not like us to linger inside too long". (Spot on!). After coffee, President Duke of Marlborough withdrew a Churchill-size cigar from an inside pocket and in relaxing fashion slid his chair back from the table. Her Royal Highness leaned towards me and whispered "Now you are anxious aren't you?".

Before a match could be struck the Duchess passed a kindly remark to husband Henry suggesting it was time to see the ploughmen. She arose with a smile. The President put away his cigar. Everyone knew it was time to go, and everyone did. Surely, a classic example of ladylike leadership.

Landrover vehicles waited to convey the VIP party first around the Competitors' plots under the guidance of the General Secretary and, secondly, around the extensive tillage machinery demonstrations and static exhibition area under the guidance of our Public Relations executive Stanley Collet. We had been briefed (goodness knows on what authority) as to where each guest must sit in the Landrovers. On no account must anyone get out and the change-over point from competition to exhibition and demonstrations must be reached prompt on time. Furthermore, not to speak to the Royals unless spoken to by the Royals.

However, in practice everything was quite different. The Landrovers were long-wheelbase type with entry door at the rear and seats along each side so that the passengers sat facing each other. As we walked to the vehicle identified by the Royal pennant, the Royal Duchess aimed towards the front door whereupon I politely asked if she would like to sit beside the driver and received the eager reply "Oh, may I?". I opened the door and she climbed aboard. The Royal Duke, the Duke and Duchess of Marlborough, Lord Macclesfield (Lord Lieutenant, Oxfordshire) and Lady Macclesfield and myself crowded into the back. Sitting facing each other, knees to knees was not conducive to seeing anything on the field however much anyone twisted and turned to peer through the windows. The situation was ridiculous. These were interested people and they were our guests. It was time to release them from their mobile cage.

"Would you rather get out and walk?"

"Yes, much better" was their chorused reply.

All disembarked and the Landrover waited on the headland. Our guests were really happy. The Duke of Gloucester took a genuine interest in both the ploughing and the ploughmen. He conversed with the competitors on the headland, discussed depth of furrow, angle of slice and soil made available. The non-speak protocol was a non-starter. He spoke of his own farm and his new combine harvester. He was in no hurry, nor were the other members of the party who had spread all over the place. When the time came to get back into the Landrover I had to tell the Duke of Marlborough that his Duchess was missing. He urged "not to worry", she would turn up when ready. There is a place and time for protocol but it is much more fun without it on a ploughed field!

After a tour of the demonstration and exhibition areas Stanley Collet brought the Royal party to the commemorative Cairn of Peace for the unveiling ceremony. His Royal Highness was assisted by three young ladies – English Margaret Kirkwood, Scottish Margery Young and Welsh Bronwen Jones each charmingly dressed in their distinctive traditional national costumes. Their job was to remove the three Union flags draped over the Cairn.

In his speech the Duke referred to himself as a farmer and said,

"Although science has assisted agriculture in many ways the art of turning the soil remains fundamental to successful farming. The life and work of the modern ploughman is vastly different from the lonely lot of his forbears, but he will always remain the craftsman of the greatest importance. His is essentially a peaceful occupation".

Trumpeters of the Regiment of Lifeguards blew a fanfare whilst each girl removed her respective flag to reveal a plinth of Cotswold stone interspersed with a built-in block brought from each world ploughman's homeland engraved with his country' name, surmounted by a model of the World Championship "Golden Plough" trophy fashioned by a Canadian blacksmith and gifted by the Ontario Plowmens Association.

Their Royal Highnesses met and conversed with all the ploughmen. The Duke, whom I felt was of a rather shy disposition, was at ease and reluctant to leave. He stretched the time by telling an amusing story of another of his numerous unveiling ceremonies. But his freedom to stay and mingle was denied by the arrival of the Rolls Royce limousine flying the Royal Pennant. A footman opened the car door and the protocol timetable was again in command.

Every ploughing match attracts the older generation of enthusiasts

Dedication of Cairn of Peace Shillingford, Oxford 1956 (Esso)

43

HRH Duchess of Gloucester admires Cairn of Peace (Latter & Son)

who evince admiration of the dedication of younger practitioners in ploughmanship. The National and World competitions in Oxfordshire were no exception. Eighty-two years old Charles Reynolds from Worcester founded the Ilford ploughing match fifty years earlier and followed with the Essex ploughing Championship and the inter-county Essex versus Hertfordshire match. He started farming at Play Street Farm, Ilford which was the Reynolds family farm for two hundred years. Then, for twenty years, farmed Fence Piece Farm at Chigwell until he retired. At Shillingford he was in his element. He had well done his share to extend the fellowship of the plough and praised the ploughing progress when he wrote his personal epilogue to a memorable event,

> "I was introduced to many people from many countries. There was a spirit of camaraderie. The old friendly rivalry; the old competitive spirit still shines and in the sparkle of an old man's memory there remain pictures taking one back through the last half century. There is a tremendous drive forward in all branches of agriculture. The horse ploughing made a picturesque scene and the medium loam soil was just right for ploughing. What more lovely sight than a pair of Suffolk Punches drawing a single furrow plough? Tractors draw two and three furrow ploughs now and a master mind of organising genius has drawn the world's ploughmen together to unite on a common footing."

44

Chapter 8
DIPLOMACY and MEDIATION

There was an unexpected supply problem when the World Contest was held at Peebles, Ohio, USA in year 1957. It might have become an item of controversy for a sensationally minded media person, but fortunately none got the idea.

American-made tractors were fuelled with petrol and vaporising oil. European tractor engines worked with diesel fuel which was not generally available in America. Some European competitors were loaned diesel engine tractors by their supportive manufacturers. Consequently, when the "diesel" competitors arrived to collect their machines for pre-Contest setting and practice they were told at the fuelling depot "We ain't got diesel", despite the advance placing of an order for a limited quantity and promise that it would be there. The provision had been overlooked, and the consequence was that the "diesel" users lost a day and a half of vital practising, whilst someone looked for their special fuel and the "petrol" users worked on merrily.

To urge some official to get a move on, it was necessary to point out that whilst American and other petrol users were making good pre-contest preparation the disadvantaged stood glumly beside their fuel-starved machines. Press reporters seeking interviews might conceive an exaggerated story of unfair treatment! The Contest Co-ordinator appreciated a tricky situation and enlisted the aid of International Harvester Company. Within a couple more precious hours the uncommon tractors were ploughing American soil.

During the pre-Contest period it was not unusual for some competitors to be visited by a fellow countryman from their national Embassy. On one occasion, in Sweden the American Ambassador thoughtfully sought leave of WPO to entertain his American ploughmen for an evening at his home. So long as absence did not conflict with the host country and WPO itinerary this was readily agreed. British diplomats were rarely seen. However, British ploughmen are important, even if the British farming community is less than three per cent of the whole population it shall not be forgotten they produce food for the other ninety-seven per cent.

At a function in honour of visiting ploughmen in Norway, Finland's lady Ambassador kindly identified several Agriculture Attachés and other members of the Diplomatic Corps present, and was surprised there was no representative from the British Embassy, for she understood all had been invited. Some time later during conversation with Minister of Agriculture Fred Peart (late Lord Peart) the suggestion that Britain's champion ploughmen at the "Ploughing Olympics" would feel happily recognised if a British diplomat could conveniently spare a little time to greet them in a foreign land on such an auspicious occasion, as do other diplomats in respect to their participating nationals.

What followed was most pleasing. When the British and European group of ploughmen arrived at Toronto Airport for the Contest in year 1963 the aircraft landed in the dark. It was difficult in the glare of terminal lights to see the faces of the official welcoming party permitted to stand in a row at the foot of the steps. But distinct from the Canadian accents were three noticeable English voices behind their three warm handshakes. They introduced themselves as British Agricultural Commissioner, British Trade Commissioner and British Consul. Their greetings were friendly and cheerful though necessarily short due to the group being quickly conducted through the arrival formalities and on to a waiting motor coach. So, our British diplomats promised to meet their ploughmen next day when they would be at the competitors' marshalling yard preparing for practice.

Along came all three and spent ample time chatting freely with ploughmen and supporters. Much pleasure was expressed for the opportunity to share their fellowship and when a few words let slip "that some time ago it was suggested to the Minister of Agriculture . . . "they cut short with a chuckle and said "Yes! We know about it! We did hear!". Fred Peart must have passed on the message!

On another occasion William Whitelaw MP (late Viscount Whitelaw) speedily found agreement with the British Foreign Office that when kept informed of the location of each World Ploughing Contest the local diplomats would be notified. Happily, diplomats do occasionally take time to share interest in the British participation and, hopefully, the Contest fixture list is with the Foreign Office.

By its constitution the World Ploughing Organisation is concerned with international relations, not in a political sense but in a grass-roots people to people way. To a relevant degree diplomacy is involved but not the common idea of diplomacy as engaged in by governments. Diplomacy can also be regarded as "cunning", whereas with WPO it has to be "mediatorship", a common friend concept as expressed in the WPO song "The Worldwide Ploughing Brotherhood" –

Far round the World this song we sing
As o'er the earth the ploughshares ring
The craft that has through ages stood
The Wordwide Ploughing Brotherhood

Chorus:
We love the land we plough so well
Each furrow mark of care can tell
On us depends our daily food
The Worldwide Ploughing Brotherhood

From every country, race and clime
We plough and sow till end of time
United by the soil so good
The Worldwide Ploughing Brotherhood
(Chorus)

Firm is the hand that grasps the plough
Warm is the heart of friendship now
Remember as you turn each sod
The Worldwide Ploughing Brotherhood
(Chorus)

An example of WPO's moderating influence is recalled in Chapter 6 of this narrative where is described how, after the eventual affiliations with WPO of both the Irish ploughing associations, the ploughmen of both north and south of the politically divided country participate in each other's matches, exchange judges, share hospitality and together share the organisation of transporting their equipment. There is an intermission where ploughmen's fellowship transcends political bigotry.

A similar WPO moderatorship was successful in Italy, too. After attending the British and World Ploughing events at Shillingford in year 1956 Italian observers invited the General Secretary to a meeting in the romantic town of Verona to discuss the establishment of a series of regional ploughing competitions culminating in a national championship event from which to select competitors for the World Contest. The promoters represented an important Italian agricultural association. They were in a hurry and wished to be immediately

47

accepted as the official WPO affiliate in Italy. Furthermore, they wished to host the 8th World Ploughing Contest and offered a venue near Rome shortly after the conclusion of the Olympic Games in year 1960. They even suggested that the WPO General Secretary would be welcome to transfer his office and abode to Verona.

After several hours of exchanges of information and discussion there was no doubt as to the enthusiasm and determination of the promoters to co-operate in the best interests of a prestigious event. But there was an unusual urgency in their desire when they announced that members of the Press were already assembled in another room to report an announcement that their particular organisation had been accepted as the affiliate of WPO in Italy!

With due respect they had to be told that whilst the WPO General Secretary had obtained a good impression and was prepared to report very favourably to the WPO Governing Board, he must first obtain the Board's agreement to the affiliation. This was not enough for the enthusiasts and the dialogue continued with refills of gritty espresso coffee. Finally, the WPO point of view was accepted with good humour and friendly understanding. And a good thing, too, because whilst visiting the headquarters of the Food and Agriculture Organisation of United Nations in Rome, where WPO was highly respected and supported by the appointment of a senior officer on the WPO Board, it was learned that there were three large, influential agricultural organisations in Italy. They respectively represented small acreage farmers, large landowners, and general farmers' co-operative trading. They were popularly known by initials, viz: Confederazione Nazionale CCDD, Confederazione Gen. Agr. Italia, and dell'UMA. Each had strong representation in Parliament. All three had different political persuasions and, so we understood, shared little, if any, collaboration.

Clearly, the eagerness of the UMA gentlemen of Verona was that they alone wanted to become the affiliate of WPO in Italy. The strategy was to compete rather than co-operate. However, interest among the country's farmers and ploughmen sparked dialogue not only among members of the three organisations but also between members of several others, UNIMA, UNACOMA, and Conacoltivatori included.

It must have become obvious that to participate in a programme of national ploughing matches linked to a series of world contests was a popular proposition among farmers and ploughmen, because for the first time the three principal organisations' Presidents conferred, namely: Dottori Conte Alfonso Gaetani (Confed. Gen. Agr.), Dottori

48

Conte Gaetani, Minister of Agriculture Rumor, Dottori Farina

Luigi Farina (UMA) and On. Dottori Paolo Bonomi (Confed. Naz. CCDD). They joined forces to form Comitato Nazionale per lo Sviluppo della Meccanizzazione Agricola for the purpose of establishing a national ploughing match and ultimately to host the World Ploughing Contest.

Soon this new found togetherness included the patronage of an honorary committee of nineteen government department Ministers headed by Minister of Agriculture Mariano Rumor, (later to become President of Italy and assassinated), and forty-five national, civic and industrial dignitaries. There was an organising committee of forty-eight representatives of relevant bodies and government departments, seven Universities, Esso Italia and FIAT Companies. Indeed, it seemed everyone was in on the act thanks to not having yielded to the plea of UMA for exclusivity during the lengthy meeting in Verona.

The 8th World Ploughing Contest was held at Tor Mancina to the north of Rome on the east bank of Tevere river on 9-10 October, 1960. Thirty-two ploughmen from sixteen countries competed. They and the members of the WPO Board were received in audience at the Vatican by Pope John XXIII, to whom was presented a full-size replica of the

49

"Golden Plough" World Championship trophy. Pope John spoke of his boyhood on his father's farm and of his brothers who persuaded their father when it was time to buy a tractor.

This was followed by a Welcome to Italy reception in the Equirinale accorded by President Giovanni Gronchi. Business meetings lasted a long, long time and were often abruptly interrupted for coffee, wine or food or to relieve one's weariness with a traditional siesta. On any aspect speakers were verbose and mostly all at the same time! It seemed an impossibility to record the minutes yet the silent lady stenographer somehow managed to get the decisions in chronological order. After each harangue there was a short pause for the benefit of translation, question and answer. Then the pandemonium would resume.

It was tempting to try a little mischief to relieve the boredom of one long, noisy session. So, under cover of much talking and gesticulating, without the Chairman intervening, a slight pencil tap on one of the two microphones centred on the table brought the patient lady secretary from her headphone listening corner of the room, tracing the cables until she elbowed her way between unnoticing committee men and inspected the microphones. Finding nothing amiss she retreated to her corner and her vigilance. Out of such rumbustious meetings there derived an unsparing generous hospitality and a World Ploughing Contest that was a memorable success.

This was the first time the final of the two-day event was held on a Sunday. Normally, the choice of dates, year, month and days is one of mutual agreement between the host country and WPO Board but the choice was not mutual on this occasion.

In both Italy and France the public choice of day for leisure and sporting activities was declared to be Sunday. Most people could be expected to attend on Sunday. But, to this idea the Ploeg Comite Nederland and the Northern Ireland Ploughing Association were opposed. They disapproved of ploughing on Sunday and withdrew their competitors from the Contest. The Northern Irish were more adamant than the Dutch because in the Netherlands committee there was a rift since not all disfavoured the idea.

However, the attendance on Sunday in Italy was about the same as it was on Saturday, but there was a religious dimension to the ploughing on the Sabbath. An Alter was erected under a canopy against the General Secretary's office, and beside the window from where Morning Mass was celebrated and a pre-Contest Service of prayer and dedication conducted by the Priests to a Congregation of competitors and spectators.

50

During the Service WPO Treasurer Philip Palmer was working at his desk beside the open window when he was observed on a couple of occasions to strike a match and reach through the window. It was a curious act and, of course, he was asked what he was doing. He explained that twice the breeze had blown out the candle light on the Alter and, since he could easily reach it, he had re-lit it!

WPO's very large flag is flown from a tall flagpole during the progress of the Contest and used as a stage backdrop during the "Golden Plough" Presentation Festival on the final evening. The provision of several smaller flags featuring the WPO "plough around the world" symbol was a generous gift from the Italian committee. These flags were hoisted at strategic points around the contest plots interspersed with the international and Italian regional flags. At the conclusion of the ploughing the large WPO flag and each competitor's national flag is already flying from poles surrounding the Cairn of Peace in readiness for the flag lowering ceremony by the ploughmen during playing of the host country National Anthem. This procedure takes care of the flags going forward to next year's host country. Meanwhile, all the other flags are lowered and collected by stewards. On this occasion not all the stewards were quick enough. Souvenir hunters had been quicker! At least two of the smaller WPO flags were missing. This was a disappointment at the time. However, a sequel to the disappearing flags incident was revealed twenty-four years later.

Chapter 9

THE LOST FLAG

Political changes are frequent in Italy and, after the 8th World Ploughing Contest at Tor Mancina in year 1960, such changes had influence on the direction of the national ploughing organisation, which was largely controlled by executives of the three founding organisations, not by grassroot members of the ploughing fraternity. Following the death of WPO's Italian Board Member Vasco Ferrari a successor was nominated but never participated. After ploughmen Cesare Sardo and Oswaldo Scaglia competed in the World Contest at Ringerike in Norway in year 1965 contact was lost, until twenty years later. During this time lapse the leaders who collaborated in Nazionale per lo Sviluppo della Meccanazione Agricola had either retired or died. A younger generation knew nothing of the great ploughing event at Tor Mancina in year 1960.

By year 1982 interest in competition ploughing revived. The Young Farmers' Federazione Italiana Clubs 3P organised a National Ploughing Festival at Piacenza on 25th July. Although the promoters were keen about an annual national championship match they had doubts, financial and other, about becoming affiliated with WPO. They were concerned that funding would be a problem yet they frequently sent groups of young farmers on visits to other countries. The cost of two airfares as prizes for the champion and runner up to go to a world contest would, by comparison, be small.

Paradoxically, a quarter of a century earlier WPO General Secretary had to resist their predecessors' plea for immediate affiliation and now he was having to persuade the new generation. Gradually, enthusiastic ploughmen aroused the interest of their organisation executives. The third of the new series of national ploughing festivals was held on the Adriatic coast at Collecorvina in July 1984 when once again your author was present by invitation to be told that Federazione Italiana Clubs 3P had firmly decided to apply for membership of WPO.

The affiliation was confirmed by the WPO Board at their next meeting and Pierfranco Faimali became WPO's Board Member in Italy. Antonio Ricci and Renzo Ronda competed in World Contest number thirty-two, in Denmark, year 1985, that was twenty-eight years after Tomaso Moreo and Nicola de Luca were the first Italians to compete in the World Ploughing Contest in 1957 in Ohio, USA.

At Collecorvino came the sequel to the disappearance of the WPO flags at Tor Mancina twenty-four years previously. Normally, the Italian match finishes an hour or so after mid-day when all the ploughmen, officials and others sit down at their celebration feast and listen to several speeches. Results are announced and trophies presented in late afternoon after the siesta. It is not unusual that the General Secretary be asked to say something whether or not he happens to know the language. If one can say only a few appropriate words in the language it is always warmly appreciated. So it behoved one to try as a matter of courtesy.

On these Italian visits Signorina Maura Arduini was assigned my helpful guide and interpreter. And so, sitting together under the shade of a tree on the side of the ploughing field, first a short speech was written in English, then translated into Italian and read aloud slowly and deliberately in the same language by the translator. Next, to be read aloud over and over again and memorised by the "student" until the translator decided the pronunciation was good enough for the occasion.

Finally, at the lunch table the tense moment came. Having apologised for the accent, congratulated the organisers of the event, the ploughmen upon their performance and praised their resumed membership in WPO, it was opportune to recall Tor Mancina and the problem of finding anyone who attended the Contest there. Thankfully the language effort was received with approving acclamation and obviously understood because a response came from two who well remembered Tor Mancina in 1960. They were there and they were supporters of competitors Giovanni Sansone and Giancarlo Castaldini. More interesting still was that their local ploughing society still commemorated the occasion at an annual autumn reunion feast when the room is decorated for the occasion with the time-honoured flags of the World Ploughing Organisation!

Thus was solved the mystery of the lost flags. In consideration of their proud and respectful preservation of the fraternal emblems of world ploughing fellowship they could surely be credited self-appointed custodians and worthy inheritors in bond to keep and treasure them for all time.

The amateur speech had a sequel, too. On leaving the function room to seek a desired siesta a crowd of eager ploughmen gathered round with an inundation of questions in Italian! The situation was bewildering. The only escape was to call for Maura who hurried to give relief. One by one she translated the questions into English and the answers into Italian. They were all about rules for ploughing. The afternoon siesta was delightfully restoring.

Chapter 10

THE PROBLEM OF SUNDAY

In the year 1961 the 9th World Ploughing Contest was held about midway between Paris and Versailles at the National School of Agriculture at Grignon, a 17th century chateau acquired by King Charles 10th in 1826 with a view to installing there the Royal Agricultural Society of France.

The WPO affiliate at the time was the Young Farmers' movement Le Cercle National Des Jeunes Agriculteurs (CNJA) whose advice and insistence was that the most popular day of the week on which to hold a prestigious event was Sunday. Therefore, the two-day Contest must occupy Saturday and Sunday. This decision was, at that time, reluctantly agreed by all affiliates of WPO except Ploeg Comite Nederland and the Northern Ireland Ploughing Association. Although the Netherlands committee decided not to enter Dutch ploughmen in the Contest they remained firmly committed to host the event in year 1962 and, thereafter, decide whether or not to withdraw from membership of WPO. However, once again there was a minority on the committee who did not disfavour ploughing on Sunday.

The Northern Ireland PA was adamant, ploughing on Sunday was completely out of the question, but their competitors could plough any other day of the week and have their plots judged at the same time as those who plough on Sunday! On Sunday their competitors could only be spectators, or not even attend.

In mediatory mood WPO Chairman Walter Feuerlein recalled there was never an original intention to plough on a Sunday since most ploughmen regarded the Sabbath as a day of rest. Vasco Ferrari (Italy) declared that Italians had no objection to holding the Contest on days other than a Sunday, but both he and Robert Perron (France) insisted it was a tradition and necessary to hold the Contest on a Sunday in order to attract spectators.

To further confuse the issue, Sweden's WPO Board Member pointed out that a Jewish ploughman came near to winning the Swedish championship and the trip to Tor Mancina. He felt sure that he could plough on Sunday but might have misgiving about ploughing on Saturday, the Jewish Sabbath! In an attempt to solve the impasse the

55

WPO Board requested the Chairman and General Secretary to explain to the dissident affiliates the French and Italian reasoning. Meanwhile, some Sabbath Day research revealed that the only day of the week which is not anybody's Sabbath anywhere in the world is probably Thursday.

As requested, a mediatory meeting with the Netherlands committee was held in Utrecht. The membership of the committee was drawn from three organisations representative of Catholics, Protestants and General Christians. This ecumenical co-operation worked well except upon the question of how to keep the Sabbath. At one stage the discussion became heated when allegations were bandied about not understanding the teachings of the Bible. Two divisions of the triple alliance were not entirely opposed to participation on Sunday but the other was adamantly opposed. Finally, the two-to-one situation was resolved by all agreeing against participation, because all three did agree it was incumbent upon them to maintain unity as a national organisation.

There was no veto on attending as spectators. Several from the Netherlands did, including two of the most vociferous objectors to ploughing on Sunday who, astonishingly, were observed to have no qualms about hastening to a Paris night-club despite the sanctity of the day!

The Northern Ireland ploughmen took advantage of their travel prize, ploughed in the stubble competition on Saturday and withdrew from the grassland competition on Sunday thus forfeiting any chance of being placed in the overall scoring.

Then the Rain Came

The deep loamy soil on the Agricultural College Farm was hard following an extremely dry summer. Sprinkler irrigation applied during two weeks prior to the Contest deposited some eight thousand tonnes of water on the competitors' plots. Then after President General Charles de Gaulle had formerly declared the Contest underway and shaken hands with every ploughman, heavy black clouds rolled across the sky and the rain poured in torrents. Everyone was soaked but the ploughmen worked to schedule. The second day, Sunday, gave excellent weather. (Does the sun shine on the righteous?). The individual standard of ploughing was commented upon as being good and an overall improvement upon previous contests.

The World Contest was confirmed as a tourist attraction. Groups of

President General Charles de Gaulle speaking at the contest in 1961

from twenty to a hundred arrived in support of their national ploughmen. The French promoters were well pleased with the attendance of farmers at a time of economic stress and political controversy.

Some days before the Contest, farmers had paraded tractors and machinery through towns and villages, blocking roads and disrupting traffic in demand for better farm prices. There were reports of several scare bombs to emphasise their ire. Whilst sitting with General De Gaulle on the ceremonial platform during the welcoming speeches there was a momentary hope that no wicked instrument was ticking beneath our seats.

The World Championship was won by William Dixon of Ontario, Canada and the Runner-up was Alan Magson of New Zealand.

Investiture

The French regard for WPO was expressed on behalf of the President and Government by investing as Chevalier du Merite Agricole the WPO Chairman Walter Feuerlein (Germany), Contest co-ordinator Robert Perron (France) and WPO General Secretary Alfred Hall "in

57

recognition of international work in the interests of better soil cultivation". Walter and Robert duly received their awards but in response to the French Government's courtesy to inform the British Government of the award to the General Secretary they received a reply from a British Minister, to the effect, that it was not appropriate to so honour a British citizen in a foreign country. Consequently, the General Secretary received from the French Government a letter apologising for the withdrawal. Walter Feuerlein was also deservedly honoured in his own country for practical and scientific work with soil cultivation with particular reference to better ploughing. At the German National Agricultural Show (Deutscher Landwirtschaft Gesellschaft) he was presented with the distinguished Max Eyth Medal by the Max Eyth Society for Improvement of Agricultural Techniques. Incidentally, at one time Max Eyth worked with the Fowler company of Leeds, Yorkshire, manufacturers of agricultural steam traction engines.

Chapter 11
UNWISE AUTHORITARIAN CHOICE

The WPO Bulletin of News and Information was introduced in year 1958 and given a free world-wide distribution every month but was gradually reduced to four then two issues annually, principally due to increased postage and paper costs. It became an important element in making contacts and keeping affiliates and their ploughmen informed of each other's activities, particularly so when items were translated from it into their own national newsletters. Its presence was the equivalent of shaking hands across oceans and frontiers as do greetings cards at Christmas. For example, in Argentina Carlos de Dios of the Instituto Inginiera Rural, was a frequent correspondent. His initiative encouraged interest in ploughing competitions with WPO rules. Only financial limitations restricted the hoped-for participation of Argentinian ploughmen in the World Contests.

In Eastern Europe, behind the "Iron Curtain," the Hungarian Agronomic Society founded the Hungarian Ploughing Organization and published an illustrated booklet on ploughing adapted to WPO rules. The first district match in Hungary was held on the Perbal State farm on 23rd March 1961. Of more than five hundred spectators most were chairmen of collective state farms, farm directors, members of scientific Institutes, Government people and political party officials.

The support of these people was such that in the course of one year five thousand ploughmen competed in one hundred and forty local ploughing matches, leading to twenty-eight regional finals from which thirty-eight finalists went forward into the National Championship at Martonvasar near Budapest. The winner and runner-up participated in the 10th World Contest on the Flevoland polder in the Netherlands on 4-5th October, 1962, using Moson ploughs and Dutra tractors.

Under the authoritarian regime the choice of equipment was made for the ploughmen and not by the ploughmen. Consequently, they were each allocated a two-furrow plough and a Dutra tractor of which there were two models. The smaller 28hp model was suitable for the two-furrow plough working to a depth of 18 to 25cm, whilst the much larger four-wheel-drive model of 65hp was for heavier work and basically designed for ploughing wider (40cm) and deeper (40-50cm) on large arable areas on upwards of 500 hectare State Production Co-operatives. Crawler versions of the same tractor were from 80hp to 100hp.

Competition in the World Contest is limited to ploughs of two and three mouldboards because it is difficult to find large enough areas where the soil is of uniform consistency and depth to accommodate multifurrow ploughs in fair competition. Logically, the smaller Dutra tractor was the proper choice to use in a two-furrow competition. The larger tractor would have been appropriate for use with a four, five or more furrow plough, but authority required that the new powerful, super tractor show its paces. It did not matter that the ploughman knew the machine was unsuitable for the job he had to do in the Contest.

At that time there were no Dutra tractors in the Netherlands and whilst attending the Martonvasar Championships the Director of the Hungarian Export/Import Agency requested WPO and the Dutch host committee to arrange with a Dutch machinery dealer to import both tractors for the Contest under special Customs licence. It was his earnest desire that the big tractor be seen at work and, of course, he was hoping for business. Regrettably, the Director was averse to any suggestion that the large tractor would attract more serious attention whilst on a demonstration area working with a multi-furrow plough or large tilling harrows rather than within the limited 2000 square metre plot for very precise competition ploughing.

The huge tractor attracted attention but many spectators felt sorry for the ploughman seated high above large wheels fitted with tyres too large to fit the required fourteen-inch (35.50cm) furrow. The two-furrow plough was diminutive in comparison with the size of the tractor and hitched close behind was out of the ploughman's sight with the controls beyond his reach. An unfortunate misuse, a failure to reveal the versatility of either tractor or plough, and a bitter disappointment for a worthy ploughman. The resultant poor ploughing by the unbalanced outfit was not a "breakthrough" for Communist industry.

Chapter 12

PROTOCOL PEOPLE and WORLD STYLE PLOUGHING

The historical importance of the plough and its development through the ages is preserved in the antique plough museum at Hohenheim Agricultural College, Stuttgart, Germany. There are models dating from the stone, bronze and iron ages. In addition to original specimens are some three hundred miniature models of ploughs manufactured in the Hohenheim plough smithy during the course of more than a century.

In this plough conscious environment the German Rural Youth Movement with the support of the Ministry of Agriculture, Hohenheim University and farmers, was hosted the 6th World Ploughing Contest on 3-4th October 1958, to coincide with the traditional Oktober Fest; an annual revel to celebrate the conclusion of harvest. Produce of all crops is displayed and tasted, especially the liquid variety. Bands, dancers, flagwaving marchers, horses and riders, all in colourful national costumes, process through the streets in parades with vehicles bedecked with fruits of the harvest and traditional tableau.

On this occasion WPO also had a place in the Stuttgart parade represented by a large globe of the world mounted on a long vehicle beside a tractor and plough surrounded by the flags of the nations participant in the World Ploughing Contest to take place on the College Farm. The parade accompanied by the world ploughmen weaved along the streets including many still in ruin from wartime bombing.

The grand opening ceremony for the Contest also had a parade of marching bands, and artistic flag-wavers to lead the thirty ploughmen driving their tractors with ploughs to the Contest plots. Behind them seated on tractor drawn wagons followed the honoured guests – Members of Government, Burgmeisters, officers of Church and State, Civic Dignitaries, University Professors. Whilst the parade was assembling four host committee members entered the General Secretary's office to ask the question "Who is the most important person on the field?" The simple answer was " The ploughmen are the most important on the field". The chorused response: "Ja! You are right, the ploughmen". "The VIPs are here because the ploughmen are here." "Genau, danke viemals".

There had been some protocol confusion as to which VIPs took precedence in the seating plan on the parade wagons. Cheerfully the four rushed away with a view of protocol from another perspective. When the parade trundled on its way the Burgmeisters, Ministers of Government and Church, Professors and Civil Servants were in smiling good humour riding high on their self-selected wooden planks and straw bales.

The quality of ploughing on both stubble and grassland was impressive. Several makes of plough produced different styles of ploughing. Most of the ploughs in the Museum portray a short, deep concave mouldboard digger variety. The shape of the ploughs used, in particular by the German competitors, were in accord with tradition simply because the soil on fertile plains and valley bottoms is mostly deep and friable and well suited for root crops such as sugar beet and vegetables.

Ploughing at optimum speed the digger plough, on stubble and fallow after roots, slices and shreds the soil to a crumbled tilth. Only a minimum of subsequent tilling is needed to make an even seedbed ready for planting, for example, potatoes, turnips, carrots, sugar beet, cabbage etc. In contrast the most important of many crops on the British Isles is grass and the management of grass in the crop rotation (particularly where soils are shallow and often heavy compared with soils on much of the plains of Europe) requires a longer and slightly convex-shaped mouldboard. This type cuts and turns a strongly profiled slice of angular shape. Grassland, also referred to as lea ploughing, was normally intended to remain in winter fallow for benefit of the natural tilling process rendered by sun, wind, rain, frost and snow. Come Springtime a stroke with harrows is usually sufficient to make a shallow crumbly seedbed to accommodate cereal seeds. Digger or mostly semidigger ploughing is adopted generally following crops other than grass for planting root crops.

Much is learned and to be learned from ploughing matches. During the first ten years of World Contests a great deal of knowledge was gleaned from observation, discussion, study of ploughing styles and types of plough associated with landscapes from whence came the ploughmen. The development of ploughing styles and types of

62

plough is influenced by environmental factors upon soils of many varieties which are subject to temperate, sub-tropical or tropical climates. The earth has dry soils and wet soils, deep and shallow soils, heavy and light according to composition in proportion of clays, silt, sands, loam, stone, chalk, peat, humus and other organic elements. The structures also relate to sub-soils of varying degree.

So many contrasts made it seem almost unlikely that ploughing matches could ever be more than local affairs in keeping with local conditions and practices. However, not to despair, the WPO fellowship in concord with plough makers and soil scientists did find a common denominator upon which to base a so-called "world style" of ploughing for the new age of factory mass-produced ploughs and powerful tractors. It is a style of ploughing adjustable to soils that vary in texture and in depths of ploughable topsoil.

Ever since the first prototype plough was a crooked branch from a tree the challenge has been to aerate soil by loosening compaction. Since centuries old rudimentary scratching of the soil, ploughmanship has developed from experience of what is provided and demanded by environment. Consequently, ploughing has adopted styles which are not universal but peculiar to specific areas. They include high-cut also known as oatseed furrow, whole furrow, digger work, semi-digger work, jointer work, chill ploughing, broken work, disc ploughing, and the extremes of both deep and shallow ploughing.

Local blacksmiths made ploughs of original design to meet the needs of local soil conditions. Not until ploughmen travelled frequently beyond their parochial boundaries, as they do now, did they become worldly aware of the differences. Until then individual opinion tended to favour one's accustomed style as the criterion upon which to judge a world ploughing championship. Such, of course was soon realised to be impracticable. Most of the styles already referred to were the scope of horse ploughmanship and superseded by the power of the internal combustion engine and mass-produced heavy ploughs.

Sound reasoning born of practical experience, laced with good humoured argument, compromised upon definition and rules for what has become known as "world style" being a fair median for competition using tractor ploughs built for a world market. This was a happy achievement considering the diverse styles spread across diverse arable landscapes around the world. Surely, there is a moral in this grass-root accordance shared by a rural vocation of so many nationalities. The same meaningful togetherness scored in Italy and in Ireland, too, as already told in chapter 8.

Austrian Competition Ploughing (1956)

Austrian Competition Ploughing (1977) – a big improvement

The Badza – tool for tilling and for harvesting

Soil preparation with hand tools in Nepal (FAO)

Sixteen Ox Team ploughing in Kenya

Horse Ploughing in Afghanistan (FAO)

67

High Cut Ploughing with aid of Presses and Boats

High-cut ploughing by Leslie Dixon, without extraneous attachments (J Hardman)

68

Semi Digger Work

Rib-Work

A World Ploughing Contest in progress (Kvernelands)

British Champion Nelson Tamblin (1962)

Chapter 13
PLOUGHING NAPOLEON'S BATTLEFIELD

The agrarian fellowship within WPO straddled the infamous "Iron Curtain" to link with ploughmen in Communist Europe. Ploughing match organisations in Hungary, Czechslovakia and East Germany affiliated with WPO in years 1962, 1964 and 1965, respectively. The Yugoslavian ploughing organisation under the auspices of Narodna Tehnika was already affiliated in 1960.

In Czechoslovakia the National Society for the Improvement of Ploughing was founded with headquarters at the Research Institute of Agricultural Engineering at Chodov, near Prague. The Society held the first national championship match in the autumn of 1963 and affiliated with WPO the following year with Dusan Hutla the first Czech member of the WPO Board. Interest in ploughing competitions grew rapidly as evinced by the fourth annual event when the attendance was thirty thousand spectators.

The educational value of ploughing competitions was recognised in Communist East Germany. On large State farms tractor driving and new age machinery operating skills were encouraged in competitions for prizes such as motor cycles, bicycles, radio sets. In similar way these east European countries were training and rewarding their athletes who ultimately achieved success in Olympic Games.

At the same time a country-wide series of local and regional matches culminating in a national championship final, and the winners going forward to the World Contest was established in Federal West Germany since year 1953. Led by Walter Feuerlein and colleagues at the Volkenrode Soil Research Station farmers were encouraged to extend their mechanical operating skills to the art of ploughmanship.

Like the rest of the nation, the ploughing fraternity of the divided German Fatherland was disunited by the "iron curtain". To visit the eastern zone was a complicated business since it was not diplomatically recognised by the western powers, nor was it regarded as a separate country by the West German Federal Republic. However, WPO Chairman Walter Feuerlein and the General Secretary did manage to cope with the formalities to allow us through the formidable death-strip of barbed wire, minefield and machine guns. This was in July of 1965 before the too-high-to-overlook concrete wall replaced the barbed wire entanglement erected in 1961, although a wall was erected in Berlin city at that time.

The entry was by railway train via Helmstedt. Stretched along the Station platform were large banners declaring "We thank our Russian Comrades for our Freedom" (roughly translated). A peep through the carriage window on the trackside revealed soldiers with machine-rifles beside every coach – doubtless to re-direct any wayward character who chose not to disembark on the platform! Being "auslanders" we had to pass through a separate checkpoint to present passport, visa and other documents. Since we were the only two we were greeted pleasantly and felt some joy emanated from our arrival, as when bird watchers espy a pair of rare specimens.

Our permit was to attend the east German Ploughing Championships at Wachau, near Leipzig. The venue was a State farm on the broad plain where the armies of Napoleon were defeated in the year 1813 by the armies of Russia, Prussia and Austria. A slab of stone commemorates the spot where Napoleon stood and watched the "Battle of Nations" as it was called.

On those same fields one hundred and fifty two years later ploughmen from Czechoslovakia, Poland, Bulgaria, Yugoslavia and Germany (east) were peacefully ploughing their separate plots alongside each other in competition under World Ploughing Contest rules. The east German ploughmen were also ploughing to choose their own champion at the most important ploughing match held in the so-called Deutsche Demokratische Republic.

Despite the inevitable rostrum from which speachyfying Government Ministers declaimed the praiseworthiness of their dictatorial regime and vilified everything and everyone to the west, outside their iron curtain, the ploughing event had a good purpose. For what purpose was the carnage Napoleon oversaw across that same plain before Leipzig? Did his bellicose ambition stretch to visualise the purpose of death and destruction was to remove whatever it was he abhorred as an obstacle in a search for a Grail called Peace? Did he ever imagine a peace when the same nationalities then warring with swords and staining the soil with their blood could, instead, work beside each other with ploughs on the same field, establishing standards of skill not to kill but to sustain the fertility of the soil for growing food that people might live? Whatever Napoleon's ambition it achieved no more than did the carnage administered by successive warring dictators.

Despite being citizens of a ruthless, dictatorial regime the international ploughmen peacefully carved the old battlefield with their ploughshares. Neat plots of straight furrows (and some not quite straight enough) lay on the plain like "landmark signatures" on a "landscape book" signifying the fellowship of Ploughmen.

After the match and the prize giving there was the usual feast and socialising. It was opportune to call a conference of representatives of the ploughing organisations and agricultural Institutes who were present from east Germany, Hungary, Czechoslovakia, Yugoslavia, Poland, Bulgaria, Romania and Russia and introduce to them the Aims and Objects of the World Ploughing Organisation.

Co-operation with similar organisations in other countries through the patronage of WPO was enthusiastically favoured, though it was recognised that financial and political problems made the desire difficult, if not impossible, for some to achieve under their present circumstances. It was agreed that invitations be extended to ploughmen to participate in international invitation classes at national events. To help achieve these ideas the late Professor Dr Kalman Lammel (Hungarian member of WPO Board) was unanimously elected to liase between WPO and east European organisations.

The east German ploughing organisation became an affiliate of WPO, re-organised and adopted the title "Komite für Leistungspflügen beim Landwirtschaftsrat der DDR" (Committee for Championship Ploughing of the Council of Agriculture for the German Democratic Republic). The west German ploughing organisation adopted the title "Deutsche Pflügerrat" (German Ploughing Council). In the fifteen years since their first ploughing match in the year 1950 (two years before joining WPO) some one hundred and twenty thousand ploughmen had participated in ploughing competitions in the Federal Republic.

At the time of writing this narrative Poland, Bulgaria and Romania ploughing fraternities had yet to become affiliates of WPO. The Baltic countries Latvia and Estonia have long histories of ploughing competitions and as soon as they were released from the Soviet yoke their national ploughing organisations became members of WPO in year 1990. In the same year the Russian Committee for Soil Cultivation also affiliated with WPO, but the membership lasted only one year until the dissolution of the Union of Soviet Socialist Republics.

Yugoslavia's first national ploughing match was held at Lublijana in

73

Plough-ladies in World Contests:

Colleen Wolf,
USA (1978)

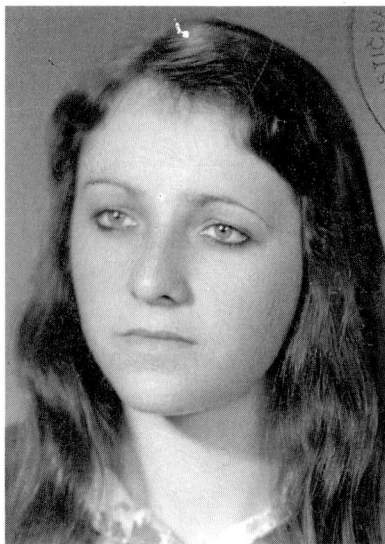

Zalokar Slavka,
Yugoslavia (1984)

Blanka Bukvic,
Yugoslavia (1990)

Helga Wielander,
Austria (1993)

Slovenia on 13-14th September, 1958 having been preceded by a series of state matches to select competitors for the event. The Food and Agriculture Organisation of United Nations was advising on agricultural development and encouraged Tony Marolt and Slavko Filipi of Narodna Tehnika Jugoslavije to be observers at the World Contests. Yugoslav competitions were judged more on mechanical knowledge and tractor driving than on ability to plough. Tests included "state of working" and "speed over an obstacle course". Out of a total of 150 points only 40 points were allowed for the quality of the ploughing. Consequently, the winners were seldom the ploughmen who ploughed the best.

The competition committee appreciated the sense of WPO rules and that what is judged at a ploughing match is the quality of the end product – purposefully ploughed land and not principally the mode of operation. Their rules were made to conform with WPO. Interest in ploughing increased, cash prizes were donated by the Government, large crowds attended the matches and everybody realised that quality and speed do not always go together. The first Yugoslav ploughmen to compete in the World Ploughing Contest were Kadovan Radanov and Ljubomir Stancic at Tor Mancina, Italy in year 1960. Also, to date, Yugoslavia is the only country to have had two lady competitors in the World Contest – Mrs Slavka Zalokar in 1984 and Miss Blanca Bucvic in 1990.

Having mentioned plough ladies, we must digress to record that the first lady competitor in the World Contest was Colleen Marie Wolf from USA in year 1978 and the one who beat all the ploughmen was World Champion Helga Wielander from Austria in 1993. To date, the world's top ploughmen have been challenged by four of the world's top plough ladies in the world series.

WPO has grown from an acorn of an idea into being like an oak tree whose branches reach across man-made frontiers to illustrate there are more naturally common interests to unite people than artificial barriers to divide them. Despite this fact, again and again fellowship shared in common interest is laid dormant whilst politicians erupt in avaricious predacity. Such happened in Yugoslavia to immerse fellow-citizens in hateful war against themselves. Consequently, Narodna Tehnika was lost from membership of WPO but, fortunately, Slovenia, where the ploughing matches began, kept out of the conflict and, having been the mainstay of the ploughing fraternity, continued the affiliation with WPO in the name of Zveza Organizacijza Teknicno Kulturo Slovenija.

Chapter 14

PLOUGH FELLOWSHIP VERSUS
POLITICAL PERVERSITY

Ploughman Bamusi Jakob Shumba was the first original national African to compete in the World Ploughing Contest. That was in the year 1965 in Norway in the 13th Contest of the series. The time was when Zimbabwe was previously known as Rhodesia; he was the runner-up to his trainer and employer Richard Light, who closely won the Rhodesian Championship. Both competed in the World Contest on the Ringerike, a fertile landscape between the folds of mountains where a stoneless medium loam of sand, silt and clay is good to plough if it is neither wet nor too dry; different from both dry sandy soil and also sun scorched, humus-lacking, non-slicing red soil they were accustomed to plough.

Participants at World Ploughing Contests, whether as competitors, visitors or in an official capacity enjoy to visit each other in their homelands when there is an opportunity. So, on one central African visit it was a pleasure to look forward to a prospective visit to the home of Bamusi. A telephone call to Richard Light's farm established that he had gone away for two weeks visiting from where ever it was he originated "out in the sticks", and there was no way of contacting him. But, news is not long delayed in Africa for within three days came a message from the farm that Bamusi was returning to greet his visitor and would be waiting at his home the following afternoon. In some way "bush telegraph" had reached him!

Whilst driving across country with WPO Board Member Eric Linnell the standard of ploughed work was noticeably good and brought forth complimentary comments, whereupon Eric explained this was "Bamusi country" where the champion ploughman coached other ploughmen who were keen to learn from him.

Bamusi's house was of rectangular shape with boarded-up windows. Apparently, most Africans prefer round-shaped houses without windows. At the entrance was an open vestibule in which stood an ornament stand displaying his ploughing trophies, WPO certificate and badge and photographs taken of him ploughing in Norway. On top of the stand were mounted photographs of Queen

Elizabeth and Prince Philip. There were three other neatly thatched round houses, one for cooking, one for tools and one for chickens. Within his compound he had two gardens, one for wet cropping and the other for dry culture. Houses and gardens were extremely clean and tidy, even the earth around the houses had been swept with a brush. All testifying to the fact that tidy ploughmen are tidy citizens.

Bamusi requested that greetings be conveyed to his world ploughing friends whom he met in Norway and to tell those who would come to Rhodesia for the 15th World Contest in year 1968 that "if Jesus is good to me I will be waiting to welcome them". As a parting gift he accepted a "Golden Plough" necktie and was emotionally grateful. With hands together in the traditional manner of prayer and "thankyou" he turned, quickly entered his house, and we left.

There was a sequel to the meeting upon returning from a visit to Johannesburg in South Africa a week later. The flight to London called at Salisbury (now Harare) where, at a pre-arranged meeting during the short stop-over, Eric Linell announced having had a visit from Bamusi Shumba who had made a long cross-country journey to deliver a live chicken for him to hand over as a present to take back to England. Eric knew the Custom control at Heath Row Airport would not welcome it so he kept it on his farm.

The spirit which prompted Bamusi's gift held a deep and sincere meaning. The live chicken represented a gift of life and reproduction. With the exchange of two gifts a necktie and a chicken two ploughmen shared the glow of fellowship.

"To Plough Well" Farms and Bus Company

During the course of a later visit to the country it was desirable to visit a typical African family farm, known as a "purchase farm" as distinct from the very small subsistence holdings and the large estates. Whilst motoring through the bush in company with WPO Treasurer Philip Palmer our driver was asked if he could take us to such a family farm. To find one he drove into a village called Gwangwadza and called at the Post Office-cum-Store to seek an address of a purchase area farmer and call him on the telephone. He was told to drive several miles to a crossroads where the farmer would wait to meet us.

Parked at the crossroads was a large American car out of which stepped a smiling African gentleman smartly dressed in khaki bush shirt and shorts. With outstretched hand in greeting Lazarus Mazenda then led us to his two neighbouring farms, totalling some four hundred acres or so, which he originally bought at a price of five pounds an acre. By a happy coincidence he had given his farm a name in the native language which translated means "To Plough Well".

The homestead, buildings and fields were immaculate. Despite the worst drought for forty years the maize crop was above average and the cattle looked in very good fettle. The farm management was obviously efficient and mechanized up to date. The family home was a comfortably large, neatly furnished bungalow. Mrs Mazenda served tea in delicate china cups and with dainty three-cornered sandwiches.

We discussed the farming enterprise and browsed through the family photograph albums particularly admiring their two small daughters, only to be shocked when the good lady said that according to tribal ritual because they were twins they were to be sacrificed. But Mrs Mazenda smilingly assured us that her twins were quite safe from any such horror because they were a precious gift from God to a happy family.

Lazarus attributed his successful crop production from the care he always took with ploughing and that is why he gave his farm the African name which means to plough well. From his profits he had bought motor buses and operated a rural service in the Ruapa district, managed by his brother. Because of his appreciation of good ploughing what better title with a moral could he give his bus company than the African one which means "To Plough Well".

"Wind of Change" in Central Africa

When news of WPO arrived in Central Africa in year 1957, Eric Linnell, a farmer at Umvukwes, and Hugh Templeton a farm machinery engineer, entered upon a campaign to improve the standard of ploughing. Contact was immediately made with WPO for information about the organisation of ploughing competitions, layout of lands, plot dimensions, rules for ploughing and method of judging. Together with colleagues they introduced ox-ploughing competitions and during the next five years, whilst in regular correspondence with WPO secretariat, graduated to tractor ploughing matches under WPO rules.

They established the Central African Ploughing Association comprising Southern Rhodesia (where the two founders were based), Northern Rhodesia (now Zambia) and Nyasaland (now Malawi). The political status of the Central African Federation lasted from 1953 until

Central Africa's first-ever Champion Ploughing Match (Wedza 1963)

1963 when it divided into three separate countries. Southern Rhodesia reverted to its status as a British colony named Rhodesia and since the ploughing association was founded in Rhodesia it was re-named the Ploughing Association of Rhodesia and carried on the good work. In five years the organisation had grown to twenty local ploughing matches, seven area finals and a national championship final.

At first disc ploughs out-numbered mouldboard ploughs by six to one, but in the national final twenty-eight competitors used discs and twenty-five used mouldboards. Competitions were also arranged for ox-ploughing and part of the final assessment was made on quality of crop yield. Another scheme which began slowly but proved very successful was the award of money prizes and medals for African ploughmen who designed ploughs particularly suited for ploughing dry land to specific dimensions. After five years of steady progress the Ploughing Association of Rhodesia affiliated with WPO and the winner of the national championship finals, Alec Philp, coached and accompanied, as his manager, by Eric Linnell, competed in the World Contest at Caledon, Ontario, Canada in year 1963. In 1965, Eric was appointed a member of the WPO Governing Board.

Meanwhile, the Colonial and British Governments were in dispute and failed to agree an independence settlement. In the year 1965 Rhodesia declared a Unilateral Declaration of Independence which the British Government declared to be illegal. Political disagreement raged between the governments and between political factions within Rhodesia. With the exception of South Africa and Portugal, the United Nations applied economic sanctions against Rhodesia. In response, Rhodesia's Government declared the country a Republic in the year 1970.

Subsequently, guerrilla warfare broke out and involved the borders of neighbouring African countries. A constitutional conference was held in Geneva during years 1976-77 but broke down. Finally, the Rhodesian Government agreed to an Election which resulted in accepting majority rule. The country's name was changed to Zimbabwe, after the name of an African civilisation, the ruins of whose buildings of the eleventh to fifteenth centuries are preserved at Zimbabwe and Khami. Likewise, the ploughing association was renamed the Ploughing Association of Zimbabwe.

Political Irritation

During the decade from 1969 to 1979 press reports about the World Ploughing Contest paid more attention to the activities of political stirrers than to concord achieved by the fellowship of peaceful ploughmen. Most United Nations countries applied sanctions with limited success against the regime in Rhodesia. The governments of Yugoslavia, Denmark, Finland, Sweden and the Netherlands went so far as to ban Rhodesian ploughmen from entering their countries to participate in the World Ploughing Contest. The British Government was equally obstinate, until strong deliberations with the Foreign Office produced a diplomatic formula. In Canada arrangements went well even when the Prime Minister took fright!

Harassment began in year 1968 when the 15th World Ploughing Contest was held on 26th-27th April on the Kent Estate at Norton in Rhodesia. Preparations were in hand from five years earlier, some two years before the country's Declaration of Independence. WPO being a non-governmental, non-political organisation did not believe its mission of international goodwill emanating from a ploughing match should be affected by political negotiations between the Governments.

However, the UDI controversy and resulting revolutionary acts which occurred gained bad publicity overseas. The Governments of Denmark, Finland, Sweden, Norway, and Republic of Ireland even dissuaded, or prevented, their own ploughmen from going to Rhodesia to participate in the 15th World Ploughing Contest. Several machinery firms were likewise prevented from sending ploughs and tractors for the competitors' use. Luckily for some the equipment was already on the way by sea when the sanction notices were delivered.

Although the Czech ploughmen were entered for the Contest despite having been advised by the British Government not to participate, they determined to go but were told that visas required for entry into the colony of Rhodesia had to be applied for in London through the British Embassy in Prague. Here is a translation from the Czech language in which the British order was issued to the Czech ploughmen:

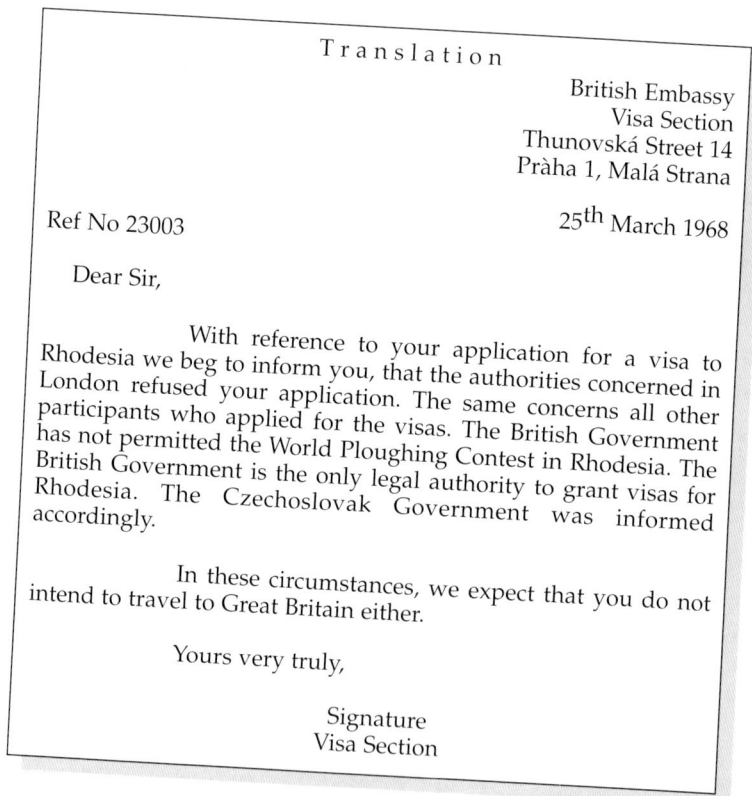

Translation

British Embassy
Visa Section
Thunovská Street 14
Pràha 1, Malá Strana

Ref No 23003   25th March 1968

Dear Sir,

With reference to your application for a visa to Rhodesia we beg to inform you, that the authorities concerned in London refused your application. The same concerns all other participants who applied for the visas. The British Government has not permitted the World Ploughing Contest in Rhodesia. The British Government is the only legal authority to grant visas for Rhodesia. The Czechoslovak Government was informed accordingly.

In these circumstances, we expect that you do not intend to travel to Great Britain either.

Yours very truly,

Signature
Visa Section

So far as the Rhodesian Government was concerned the World Contest ploughmen from any country had no need to apply to any Embassy for a visa to enter Rhodesia. The Rhodesian Department of

Immigration required and received a complete list of the names of the international ploughmen and accompanying members of the WPO group together with flight arrival times whether travelling with the group via London or directly from elsewhere. For them the need for visas was waived by the Immigration Department in Salisbury (now Harare) as notified in the Order dated 7th March 1968,

> "In view of the circumstances surrounding their visit to this country, it has been agreed that special dispensation be granted to them, and therefore, there will be no need for them to obtain Temporary Permits on arrival, the fee for which is normally £2, and they will all be admitted on Visitors Entry Certificates without further formality or charge."
>
> D A GORDON
> for: CHIEF IMMIGRATION OFFICER

Mr Geoffrey Hirst MP raised the matter in the British House of Commons:

AN EARLY DAY
129 *LOAN OF PLOUGHS FOR WORLD CHAMPIONSHIPS*
Mr Geoffrey Hirst
That this House considers the action of Her Majesty's Government in refusing to grant a temporary export licence to Messrs Ransomes Sims and Jefferies Limited to cover the loan of their ploughs for the World Ploughing Championships to be held this year in Rhodesia not only absurd and petty in nature but harmful to British export interests by destroying a world wide opportunity of advertising British made prestige agricultural equipment.

Lord Inglewood asked a leading question in the House of Lords which brought about the debate recorded in Hansard:

HOUSE OF LORDS
Tuesday, 6th February, 1968

The House met at half past two of the clock (*Prayers having been read earlier at the Judicial Sitting by the Lord Bishop of Derby*):
The LORD CHANCELLOR on the Woolsack.
The Lord Congleton – Sat first in Parliament after the death of his brother.
RHODESIA: SANCTIONS AND PLOUGHING COMPETITION
Lord INGLEWOOD: My Lords, I beg leave to ask the Question which stands in my name on the Order Paper.

[The Question was as follows:

To ask Her Majesty's Government whether they will reconsider their refusal to allow British competitors in the world ploughing competitions to take their ploughs with them to Rhodesia where, in pursuance of arrangements made five years ago, the world ploughing championships will be taking place in April, 1968.]

Lord BESWICK: No, my Lords. Exports of agricultural machinery to Rhodesia from this country have been prohibited since February, 1966, because of the declaration of illegal independence. This policy has been agreed by both Houses of Parliament and it would be wrong to make exceptions.

Lord INGLEWOOD: My Lords, may I ask the noble Lord whether he realises that this high moral line on this quite small issue shows a complete lack of understanding; that this is not a question of agricultural exports, but a question of two ploughs being taken to Rhodesia in the interests of this country and brought out again after something like 48 hours? Does he not appreciate that this world ploughing competition is something quite different from international football matches, where large crowds are attracted and there is wide Press publicity, and that nothing but good will can flow from furthering competitions of this kind? Surely the noble Lord cannot say that there is a surplus of good will in the world to-day.

Lord BESWICK: My Lords, the noble Lord has expressed an opinion, but it is not one that is shared by others. The fact is that other organisation and other people have accepted the advice of Her Majesty's Government not to go to Rhodesia for these international events in order to enable themselves to be exploited by the Rhodesians for propaganda purposes, and I see no reason why an exception should be made in this case.

Lord INGLEWOOD: My Lords, does the noble Lord's reply confirm a rumour that I have heard: that Her Majesty's Government have been in contact with other countries trying to wreck this competition?

Lord BESWICK: No, my Lords. What the noble Lord has possibly heard is that Her Majesty's Government themselves honour their obligations in this matter, and have asked that other countries do the same.

Baroness EMMET of AMBERLEY: My Lords, because of the immense benefit that would result to the native population of 400 million Africans, in whom we all take the greatest interest, in seeing modern methods of agriculture, has the noble Lord never heard of the saying about "beating swords into ploughshares"?

Lord BESWICK: Yes: I do happen to have heard of this saying. It is a pity that Mr Smith has not heard of it.

Lord BLYTON: My Lords, is my noble friend aware that Rhodesia

84

was chosen as the venue for the ploughing competition five years ago, before the present issues arose between us and them? I support the Government in its Rhodesian policy, but are we not carrying it too far now in stopping two ploughs from going in for an international competition that will result, if we win, in giving Britain the exports which since devaluation we particularly need?

Lord BESWICK: No, my Lords. The prohibition is on the export of agricultural machinery. If an exception is made in this case, there is no doubt at all that the Smith régime will exploit it for propaganda purposes.

The Earl of MANSFIELD: My Lords, arising out of the original reply, I would ask: Do Her Majesty's Government not realise that this competition has a considerable amount of importance in international agricultural circles, and that the winning of it will greatly add to the prestige of Britain? Do they further realise that Scottish and English competitors taking part will be gravely handicapped if they have to use ploughs to which they are not accustomed? Surely, as has already been pointed out, this is not an export question at all? These machines are in for but few hours. Surely the Government will not persist in harming our own competitors' chances, and still further damage our Rhodesian relations by insistence upon this childish spite?

Lord BESWICK: My Lords, I can understand the feeling of some noble Lords. I myself should like to see British people going there with British ploughs and taking part in a competition of this kind. But the way to enable that sort of thing to take place is for the Rhodesian régime to return to constitutional rule.

Earl FERRERS: My Lords, is the noble Lord aware that in order to justify their embargo on the use of a British plough, and to see that British interests are not prejudiced by other competitors taking part, the Board of Trade have in fact written to Norway, Sweden, Finland, Belgium, the Netherlands, Ireland, the United States and Canada, asking them also to boycott this international competition?

Lord BESWICK: My Lords, I have already stated that Her Majesty's Government take seriously this question of effectively implementing their undertaking to apply sanctions to Rhodesia.

Lord MITCHINSON: My Lords, is it not the view of the Government that they ought to do what Parliament has bidden them to do, and not the opposite?

Baroness HORSBURGH: My Lords, can the noble Lord say whether all the ability in the Government cannot think of some means by which, for 48 hours, these two ploughs can be looked after and be brought back, so that they do not count as exports?

Lord BESWICK: My Lords, as a matter of fact, it is not a question of two ploughs; there are more than two ploughs. A number of the other countries are in tending to use British ploughs. They have been told, and they have agreed, that it would be wrong in this case to break the policy of Her Majesty's Government which is to prohibit the export of agricultural machinery to Rhodesia.

Lord GRIMSTON of WESTBURY: My Lords, why do the Government wish to make it more difficult for British competitors to win this ploughing championship? And does it not make a nonsense of their assuming to back the "Back Britain" campaign?

Lord BESWICK: My Lords, what I think makes a nonsense is for the noble Lord and other noble Lords to get up in the House, having agreed, as this House has done, to apply sanctions to Rhodesia, and to make fun of any attempt to apply them.

Lord CONSFORD: My Lords, from the Minister's earlier answer, did I not correctly understand him to say that the object of this was to prevent this ploughing competition from taking place in Rhodesia, but is it not a fact that it does not have that effect at all? There is nothing to stop these British competitors from lawfully going there to compete. The only question is with what machine they compete. Is that not the point?

Lord BESWICK: My Lords, that is perfectly true; and if a competitor went to Rhodesia and had a British plough which he found in Rhodesia, and used it there, it would be perfectly lawful. What would not be lawful would be for anybody to export agricultural machinery to Rhodesia.

Lord BALERNO: My Lords, does the noble Lord realise that this decision is causing considerable resentment in rural Scotland, where many relatives of the rural Scottish in Rhodesia exercise a moderate influence on that Government?

Lord BESWICK: My Lords, I have no doubt that this decision does cause some resentment in some parts of rural Scotland. Equally, I have no doubt that a decision to reverse our policy would cause a good deal of resentment in parts of Africa.

Earl FERRERS: My Lords, while, of course, the embargo is on exports, how can one of these ploughs possibly be classified as an export when it is going to be brought back again?

Lord BESWICK: My Lords, the noble Earl has more information than I have on this. As a matter of fact, an application has been made for an export licence which has been refused.

The LORD PRIVY SEAL (Lord Shackleton): My Lords, if I may intervene at this stage, I had hoped that the noble Lord, Lord Airedale, who for a long time has, I know, been trying to ask a question, would have a chance. I am rather conscious of the thermal discharge from the television lights and it seems to have been added to here. I do not know whether your Lordships wish to go on pursuing this matter, but my view is that perhaps we might go on to the next Question, which the noble Lord, Lord St Oswald, is going to ask.

Lord AIREDALE: My Lords, I am much obliged to the Leader of the

House. If one question from the Liberal Benches would be permitted, may I just ask whether if a British violinist proposed to go to Rhodesia to give a recital in Salisbury, this would be banned as being an export of violins to Rhodesia?

Lord BESWICK: My Lords, the noble Lord possibly thinks that is funny. It is always easy to poke fun at anybody trying to apply a principle. What the noble Lord should do is to consult the remainder of his colleagues on the Liberal Benches and ask whether or not they are in favour of keeping up these sanctions against the illegal régime in Rhodesia.

Lord BYERS: My Lords, may I ask the Minister whether he is aware that our policy on this matter is absolutely clear?

Despite the political discord there was keen interest in the ploughing fraternity and among people with relations in Rhodesia to join a common interest travel group to attend the Contest. An itinerary was arranged with South African Airways as official carrier. In fact, the international World Ploughing group of competitors, officials and supporters was the very first common interest group to fly with South African Airways.

As the day approached for the British and European ploughing group to foregather at London Heath Row Airport political pressure to call off the Contest increased. South African Airways UK Sales Manager Keith Rennie grew increasingly anxious lest his aeroplane left without any passengers. When the Boeing 707 did lift off the runway and disappear into clouds, with the group of one hundred and ten WPO passengers on board, there was no person on London Airport more relieved and happy than he.

Because of an international embargo denying SAA the use of air navigational aids to fly the direct route over the continent of Africa the aircraft took an oversea route along the coast of Portugal, around the west of Africa and southwards to eventually turn east across Angola and land at Salisbury (Harare) airport in the pleasant climate of the high veldt. The round-about-flight took two and a half hours longer than the normal direct route. During the night navigation employed the use of a sextant housed in a dome on top of the fuselage.

Air travel made the world feel smaller and brought distant ploughmen and friends together as neighbours. They can ride the sky to where on earth there are other ploughmen to welcome them and other fields to plough. Perhaps, like the furrow from the peaceful plough, the vapour trail from the high-flying jet is also a symbolic reminder of our unity.

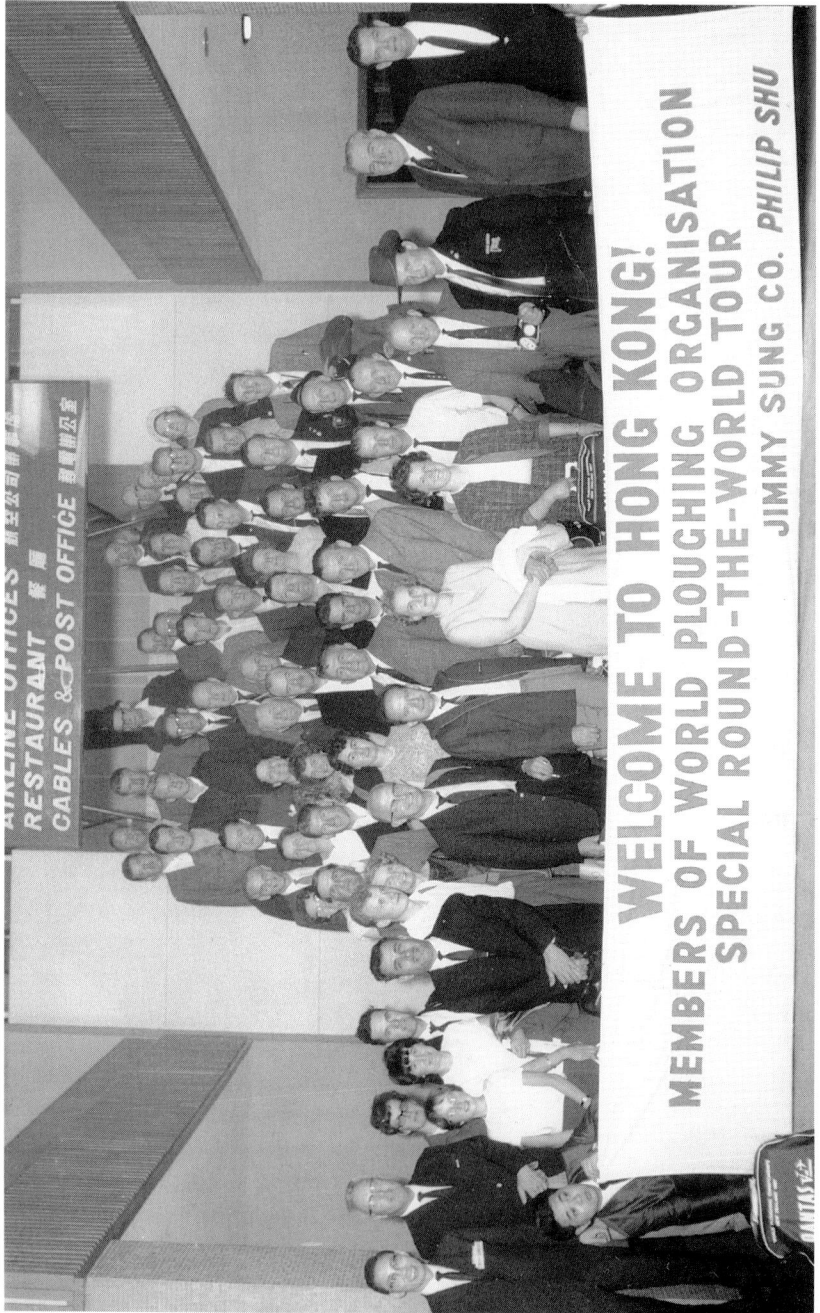

WELCOME TO HONG KONG!
MEMBERS OF WORLD PLOUGHING ORGANISATION
SPECIAL ROUND-THE-WORLD TOUR
JIMMY SUNG CO. PHILIP SHU

Chapter 15

WELCOME TO RHODESIA

There was a warm welcome upon disembarking from the aircraft. Formalities were dispensed with whilst a team of porters immediately transferred the group's baggage directly into waiting coaches. One porter brought a message from the Baggage Master requesting that passengers from Cumberland should delay boarding the coaches until he had met them. The Cumbrians stood aside and waited. When the transfer was complete Baggage Master Joe Hodgson emerged smiling to greet fellow Cumbrians! Joe originated from Broughton, near Cockermouth, and was the Uncle of Rhodesia's Prime Minister Ian Smith.

The group settled into their headquarters Ambassador Hotel, and on their first evening Bamusi Jakob Shumba arrived to greet his world ploughing friends, as he said he would, and join them for dinner.

To the relief of the competitors the ploughs had arrived in advance and the tractors specified by the ploughmen were all available, including a number of FIAT tractors of completely new design. Surprisingly, despite UN sanctions, the Italian company chose to introduce their new product to the world at the World Ploughing Contest in "isolated" Rhodesia. From that event it was successfully launched on the world market.

Unfortunately, of the seventeen nationalities originally intended to participate in the Contest two, Czechoslovakia and Hungary, were unable to confirm their entries because of the visa problem, five nationalities entered and were obliged to withdraw, leaving ten nationalities represented – Great Britain, Netherlands, West Germany, New Zealand, USA, Canada, Northern Ireland, Belgium, Australia, Rhodesia.

Ploughing the deep, red soil was a problem. Although the surrounding natural vegetation of trees and bush was vigorous, the cleared arable area lacked humus. This happens when red soils containing iron oxide are cleared and laid fallow to grow crops. The humus in the topsoil is decomposed by exposure to the severe African sunlight. When put to the plough the soil does not slide from the steel mouldboards but cakes on to them. At the end of each furrow the competitors had to scrape free and polish their mouldboards prior to the next run. This was not only time consuming but de-energising work in the burning heat of tropical sunshine. So much so, that Adolf Preuss, from Germany, collapsed and was carried from his tractor to rest under the shade of the First Aid tent during the interval reserved for judging of the "opening splits". After careful nursing he recovered and ploughed to win "Runner-up" only four points behind winner of the World Championship Marinus Schoonen, from Belgium. Englishman Ken Chappell was placed joint-third with Germany's Heinz Henkin, followed by Scotland's Robert Anderson placed fifth.

For most of the ploughmen this first tropical experience was a big surprise. Temperate and wet land farmers are accustomed to the description "sticky when wet" and never expect to find a soil "sticky when parched"!

However, there was one mouldboard that did clean under these conditions. This research model was not used in the Contest but was demonstrated by a Japanese manufacturer who, apparently, had the same disregard for politically imposed sanctions as did the people from FIAT. His new plough was fitted with plasticised mouldboards from which the humus-lacking soil slid away without caking, except upon one portion of less than a square inch where the flat head of a steel holding bolt was flush with the plastic surface. Here the bothersome soil clung like a limpet – a knob on an otherwise soil-free board.

Japanese and Swiss were early users of plastic mouldboards, since when they have become popular and were used by several ploughmen in the 42nd World Contest in equatorial Kenya in year 1995.

The art of farming where the sun blazes so strong is not to leave cleared land exposed too long otherwise within a few years the soil may become barren. The system is, and the climate allows, round the year cropping. Seasonal changes are minimal compared with those of other climates, although there are periods of heavy rainfall and periods of draught. Good farming practice ensures the land is almost continually covered in crop thus winning two rewards, protecting the humus from breakdown by strong sunlight and providing prolific all year production

Isolated

That we were holding a prestigious international event in a country frowned on by the United Nations, politically isolated and cornered by

Alfred Hall, Eric Linnell, Sir Humphrey Gibbs, Walter Feuerlein

The author with Prime Minister Ian and Mrs Smith

economic sanctions was to us, the visitors, not noticeable. The World Ploughing Contest was a festival which all the communities happily shared – incorporating the serious work of ploughing with its educational and sporting purpose, exhibitions and demonstrations of modern farm machinery, African art and culture including dance and song, wildlife safari, family visits and banquets.

The only political limitation went unnoticed by the thousands of spectators. It was a restriction by the British Government upon Her Majesty's Governor General Sir Humphrey Gibbs from being associated with the event. Like every other levelheaded person Sir Humphrey admired the aims and objects of WPO but he was obliged to stay away on the public days. Nevertheless, he did his British duty by visiting privately during the pre-Contest preparation to meet and observe the prowess of his fellow British subjects on their practice plots, as well as by warmly receiving the British and Commonwealth members of WPO Governing Board at his residence.

Prime Minister Ian Smith, farmer and RAF wartime pilot, was in attendance during the course of the Contest. He presented the "Golden Plough" trophy to new World Champion Marinus Schoonen, and the "Silver Rosebowl" to Runner-up Adolf Preuss at the Awards Ceremony in the Salisbury (Harare) Sports Stadium. He also surprised the Italian ploughmen by speaking to them in their own language which he had learned from a protective Italian family whilst an escaped prisoner of war from Italian custody!

For the first time in history champion ploughmen from ten countries ventured, in friendly rivalry, to test their ploughmanship on an unfamiliar, problematic, tropical soil. After many years of collaborating with Rhodesian farmers to promote a better understanding of the fundamental importance of good ploughing, WPO did not draw back from its mission but proceeded with its Aims and Objects, despite the country developing into a political hot-spot. Those who participated in the Contest enjoyed great hospitality and did far more to educate and encourage fellowship and understanding than did those who were forced to opt out.

For the next ten years feuding politicians allowed their dispute to stalk every World Ploughing Contest. (Likewise they interfered with the Olympic Games). But, the fellowship of the ploughing fraternity held firm. Perhaps, the modern term for the spirit of solidarity is "soul-power"?

The same organising/host committee that worked so well for the Contest in one-time Rhodesia in year 1968 continued the lessons from the ploughing on Kent Estate and were again organising/host committee for the 30th World Ploughing Contest held on the lighter, sandy land of Brechin Estate in May, 1983 when the regime had changed from being Rhodesia to become the new Zimbabwe.

Chapter 16

PLOUGHING into POLITICS
in YUGOSLAVIA and in BRITAIN

In Yugoslavia the host organisation was Narodna Tehnika, a union of cultural organisations consisting of seventeen specialised societies including the Yugoslav ploughing Championships Committee affiliate of WPO. The 16th World Ploughing Contest was held on heavy, peaty, black soil of the estate called "Beograd Kombinat", an enormous agricultural-industrial estate of more than 47,000 hectares (116,237 acres) bordered by the Danube and Tamis rivers and the Karas canal. The dates were 26-27th September, 1969.

The host committee was embarrassed by the Yugoslav Government's refusal to permit Rhodesia's champion ploughmen participate in the Contest. All other nationals were welcome as competitors. They came from nineteen countries and included the new crop of champions from countries which had boycotted the previous years' Contest in Rhodesia. However, the Government was at pains to point out that the ban was a ruling of the United Nations Organisation, of which Yugoslavia was a member.

A preparatory visit confirmed the eager intentions of the host committee to stage an exceptionally good contest and prepare a warm welcome for participants and visitors alike. There was a real sympathy for any who, through no fault of their own, were to be denied probably their one chance in a lifetime to compete in a world contest and, indeed, have such a golden opportunity to visit another country as honoured guests.

The President of Narodna Tehnika was Todor Vijasinovic, one-time secretary of Parliament and a close associate of President Tito. We spent much time together discussing the basic rights of ploughmen not involved in politics but adversely affected by other people's political disputes. We took a long tour from Zagreb in the north to Belgrade and south to picturesque Mostar and its glass-house vegetable and flower production, to Skopje in Macedonia where the water shortage was such that river beds were parched dry and where, at one place, it was possible to take a short cut by driving the car along a dried up river bed.

95

Todor understood perfectly well that the debarred ploughmen were the victims of an international political dispute in which they had no part. However, the situation was delicate and Yugoslavia was obliged to abide by the UN sanctions in all aspects. He had discussed the matter with President Tito and the consequence was they could welcome the arrival of Rhodesians holding British passports as visitors only, but definitely not as competitors. When asked if he, as President of the host committee, would write a simple letter to WPO's Board Member, Eric Linnell, in Rhodesia, explaining the unfortunate reasons that debarred the ploughmen from competing, but also saying they would be welcome as non-participant, visiting British passport holders. His answer to me was "No. You write the letter and I will sign it". This we did. The letter was posted in Belgrade to Eric in Rhodesia.

Todor Vijasinovic presents substantial prizes

This was the letter:

"Dear Mr Linnell

It is with real regret that we have to inform you of the following. Despite the fact that the entries of the Rhodesian ploughmen are accepted by the World Ploughing Organisation, and contrary to the wishes of the Yugoslav National Ploughing Committee and of the Yugoslav ploughmen, our organisation is not permitted to invite and give hospitality to the ploughmen from Rhodesia. We have discussed this problem and all the circumstances with the high authorities in Yugoslavia who have advised us of their standpoint in the matter.

It is that, as you know, the Security Council of the United Nations adopted a Resolution to apply obligatory sanctions against Rhodesia and, we are advised that, as a member country Yugoslavia is obliged to abide by this Resolution. Consequently, at the beginning of this year our Federal Assembly introduced a law which forbids having any economic relations with Rhodesia and also forbids the entry into, or transit through, Yugoslavia of any person with a Rhodesian passport.

We have asked our authorities if there is a way to overcome these restrictions so far as the Rhodesian ploughmen are concerned in order that they may participate in the sixteenth World Ploughing Contest. As you know, in Yugoslavia we like to extend a warm hearted welcome to people from all parts of the world.

However, we are very sorry to be told this is quite impossible under the present political situation. It is something we did not expect when it was originally decided to hold the World Ploughing Contest in Yugoslavia. We ask you to please understand that this ruling does not diminish the friendly feeling of fellowship which Yugoslav ploughmen have for other ploughmen in all parts of the world. Please convey our regrets and our personal greetings to the disappointed ploughmen Smith and Shumba and also the ploughmen of Rhodesia and the Ploughing Association of Rhodesia.

Yours sincerely

Todor Vujasinovic
Narodna Tehnika Jugoslavije"

My next request was to seek an assurance that when the Rhodesian observers arrived at the immigration and customs examination desks the forces of burocracy would not turn them away. To alleviate this happening Todor Vujasinovic introduced me to the Chief Immigration and Customs Officers who kindly agreed that I stand with them and indicate the Rhodesians as they approached the examination desk. The queue from aircraft shuffled forward and I saw Eric and his colleagues looking anxious as they entered the building. I waved to him from my privileged position and gave the thumbs up sign. As soon as they arrived at the examination point passports were examined and they were waved through together with their luggage. From then onwards they enjoyed every facility, every aspect of hospitality and every courtesy. Eric performed his duties as a Member of the Governing Board but, of course, the champion ploughmen of Rhodesia had been left at home.

The winner was Flemming Thyssen of Denmark with Peter Anderson from Australia the Runner-up. The same political restriction was applied against the Rhodesian ploughmen at the following year's contest which was otherwise a very successful event in Denmark.

Problems in Britain

In the meantime, some ambitious planning was in preparation for the eighteenth World Ploughing Contest to be held at Wellington in Somerset in the autumn of the year 1971. The British Ploughing Association was anxious to stage a prestigious international event and welcome all participants of whatever race, creed, religion or nationality regardless of the politics it was their fortune or misfortune to live with. Since the Yugoslav and Danish governments had discriminated against the ploughmen from Rhodesia WPO was anxious to seek an assurance that the British government and it's delegated authorities would not deny any WPO ploughmen entry into the United Kingdom.

In year 1956 ploughmen from East Germany had met with alleged difficulties at the Allied Travel Office in Berlin. These difficulties turned out to be not the fault of the Allied Control as claimed by the Communist authorities, but by the Communist last minute refusals to confirm exit permits. Since both East German and Rhodesian passports were not recognised by the United Nations we needed to know whether innocent ploughmen from two outlawed regimes would be treated any different from all other competitors.

Well in advance of the Contest in England I sought the advice of Fred Peart MP who became Minister of Agriculture and ultimately Lord Peart. He passed my enquiry to Merlyn Rees, Home Secretary, whose reply was:-

"Travellers from East Germany require visas and these can be issued only on travel documents issued by the Allied Travel Office in Berlin. This is because the Government does not recognise East Germany as a state, and in consequence does not accept East German passports for travel and visa purposes.

As regards contestants from Rhodesia their admission would depend, in large measure upon their holding passports which are acceptable to Her Majesty's Government. In conformity with the Resolution adopted by the United Nations Security Council in May 1968, all Rhodesian passports have been declared invalid, and the holders of such passports who are not also citizens of the United Kingdom and Colonies are refused admission to the United Kingdom except where there are strong compassionate circumstances to justify a visit."

This reply was all right so far as it went but it was yet too early to know who would be the winners of the 1970 Rhodesian Championships. How to obtain a British passport in Rhodesia? We could not know whether the two ploughmen would be holders of passports either British or Rhodesian or of any. A Rhodesian competitor of European decent might hold a British passport but a competitor of African decent might hold a Rhodesian passport. Were we to assume that the British Authorities would discriminate against the African ploughman and deprive him of his well-earned honour to compete in the Contest in England? Much depended upon the answer to this question. It was a touch-and-go decision whether under the circumstances, the Contest should be held in Great Britain. The answer came from the Home Secretary as follows:

"The entry to the United Kingdom of ploughmen from Rhodesia would depend upon their being in possession of valid United Kingdom passports, and this in turn would depend in the normal course upon their being citizens of the United Kingdom and Colonies, British subjects without citizenship, or British protected persons. Holders of other travel documents acceptable to the government may also be admitted, but, as I explained in my earlier letter holders of Rhodesian documents are not eligible for admission. I should emphasise that this is not what Mr Hall calls discrimination; the United Kingdom is bound, in common with other members of the United Nations, to apply sanctions against the illegal regime in Rhodesia, and this involves refusing to admit to this country people holding Rhodesian travel documents".

What we hoped would be a simple resolution turned out to be nothing of the sort. This was only the beginning of a protracted correspondence including letters to several hundred Members in the House of Commons and House of Lords. In the midst of it all there was a general election followed by a change of government from Labour to Conservative. Fortunately, we had kept the senior members and other MPs of both parties fully informed of the proposal to hold the World Ploughing Contest in England and of our need for an assurance that ploughmen would not be denied entry into the United Kingdom to compete.

Subsequent to the change of government the question was referred to Sir Alec Douglas Home. The Rhodesian Ploughing Championships had been held and the two winning ploughmen were Samuel Koti, a thirty-three year old African-Rhodesian farm labourer on a seven thousand acre farm, and Richard Boswell, a thirty-six year old Rhodesian farmer of two hundred and fifty acres who was born at Norwich in England. Koti held a Rhodesian passport and Boswell held a British passport.

Question in the House of Lords

A letter dated 24th August 1970 received from the Foreign and Commonwealth Office, signed by Lord Lothian said, "I confirm that the position explained by the Home Office in February and March this year has not changed in relation to the entry into the United Kingdom of Rhodesian contestants in the eighteenth World Ploughing Contest in 1971". The attitudes of both Labour and Conservative Governments were the same. By January 1971 Lord Lothian again advised "I can confirm that the position has not changed since I wrote to you on 24th August last. Entry to the United Kingdom will not be denied to a Rhodesian ploughman who is in possession of a valid travel document acceptable to the British Government; including of course, a United Kingdom passport. As the Home Office explained to you last February and March, holders of Rhodesian documents could not be admitted".

It was now July, 1971, and all entries for the eighteenth Contest had been received including those of the two Rhodesians. The big question was would they be refused entry into the country? From Rhodesia applications were submitted to the British Vice Consul in Pretoria, South Africa whose office was dealing with the issue of passports and

100

travel documents. The response to the applications was again disappointing.

The Rhodesian Political Department at the British Foreign and Commonwealth Office told Boswell that his eligibility to hold a UK passport was withdrawn and that he was denied entry to the United Kingdom. Likewise, Koti was informed from the British Office in Pretoria that he could not be given travel documents to visit the United Kingdom. The Earl of Mansfield now raised the subject in the House of Lords as recorded in Hansard

"The Earl of Mansfield asked Her Majesty's Government:-

Whether (1) they are aware that the British passport held by Richard Boswell an intending competitor in the eighteenth World Ploughing Contest to be held in Somerset in October, has been withdrawn and that he is not to be allowed entry into Great Britain which is his birthplace and family homeland; (2) they are also aware that they are refusing to allow a Rhodesian African – Samuel Koti – the Rhodesian championship ploughman, permission to come to Britain to compete in the Contest; (3) in view of the delicacy of the negotiations now proceeding between this country and Rhodesia and the urgent need for a settlement of the position between the two countries, they realise the harm that is being done by these unnecessary pinpricks; and (4) they will reverse the decisions alluded to."

The Parliamentary Under Secretary of State, Foreign and Commonwealth Office (The Marquess of Lothian):-

"Her Majesty's Government's policy is to seek a just and sensible settlement of the Rhodesian problem which will be in the best interests of all Rhodesians and which will enable Rhodesia to take her place in the international community. In pursuit of this aim, we are currently engaged in exploratory exchanges with the Rhodesians to try to find an acceptable basis for negotiations within the ambit of the Five Principles. Meanwhile, sanctions must continue. In this situation there are objections to a representative Rhodesian team participating in an international competition in this country, but we are in touch with the organising body in order to see whether there is any way in which these objections can be met."

By this time several members in both Houses of Parliament were describing the refusals as silly. After all, no denial of entry to the UK had been placed for example, upon tennis players and golfers. This gave the clue to the discrimination against the ploughmen. It was embodied in the use of the word "team". Unfortunately, when two ploughmen travel together to participate in a ploughing match they are often spoken of as a "team" whether or not they are from the same village or county or national ploughing match in which they both qualified to participate in a World Contest. The fact is, of course, that ploughmen compete as individuals. Each is determined to out-plough the other just as much as an individual golfer or tennis player is determined to do better than all the other golfers or tennis players.

Naturally, they are good companions, but on the field they are in competition with each other. If they were members of a team, as in football or cricket, then they would operate the same plough and tractor together and plough the same furrow. One ploughman might drive the tractor and the other control the plough. But, that is not the case. A ploughman in the World Ploughing Contest is entirely on his own, without help or guidance from any other person. His ploughing is all his own work just the same as it is whilst ploughing a field alone on the home farm.

Tactless

At this very time of intense effort to educate the Civil Service in aspects of the individual art of ploughmanship an impatient member of the Ploughing Association of Rhodesia took it upon himself to submit Samuel Koti's application for a passport to the British Embassy in Pretoria, South Africa. In his supporting letter he said that "Mr Koti is now eligible to represent Rhodesia in the World Ploughing Contest to be held in England".

The Foreign and Commonwealth Office in London threw up their hands in horror at the phrase "represent Rhodesia". Probably in glee too, at the tactless action. My immediate response was to telegraph the Ploughing Association of Rhodesia in strong terms to be factual in their expressions and to refrain from interfering whilst negotiations were being conducted in London. This was a set back. However, a brisk correspondence was passing to and fro between WPO and a number of members of both Houses of Parliament who were tendering good advice whilst lobbying the Ministers. A key word was introduced by Earl Bathurst when he said the ploughmen were "from" Rhodesia which did not mean they represented Rhodesia.

The Ploughing Association of Rhodesia had kicked the ball between their own posts and scored for the British Government. The result was that Lord Lothian wrote to say that he had believed from my earlier correspondence that "the Rhodesian ploughmen would be competing as individuals and not as a national team" but that it was apparent from the PAR letter to the Embassy in Pretoria that this was not the case. The "letter clearly indicates that the ploughmen are regarded as representing Rhodesia thus creating a completely different situation. Although we wish to foster the growth of multi-racial activities in Rhodesia we cannot do so in circumstances in which this might be regarded as a contravention of Security Council Resolutions. I regret therefore that it will not be possible to provide a travel document to Mr Koti to participate in the World Ploughing Contest in Somerset. For the same reason I can no longer give you an assurance that we can allow entry to this country by Mr Boswell or by any other person purporting to represent Rhodesia at the Contest." My reply was as follows:-

"Dear Lord Lothian

Your letter of 21st June 1971 is an utter disappointment and a complete letdown. Your statement that you would be rendering assistance to an illegal regime by facilitating the presence of a Rhodesian 'team' of ploughmen 'in an international competition in this country' is quite unacceptable.

The so-called 'team' is a misnomer. The two ploughmen of Rhodesia who were first and second place winners in their national ploughing match have no more to do with, and no more support from, the Rhodesian Government than have the two similar British ploughmen to do with, or enjoy support from, the British Government which has never in any way helped the British Ploughing Match or any of the British ploughmen who have been fortunate enough to win it and participate in previous World Ploughing Contests.

We use the term 'team' purely and simply because the two ploughmen are colleagues and for you to infer that the use of the term 'team' implies being official representatives of a government or political movement is not only nonsense but dishonest. Obviously, since you make this an excuse it is unfortunate that we ourselves ever described our competitors as 'team-mates'. Every affiliate ploughing organisation of WPO is invited to enter two competitors but some only send one. They compete absolutely as individuals for the honour of being declared World Champion. They do not compete as a team. There is no team prize. There is a trophy for the World Contest winner and a trophy for the runner-up. Ploughmen in competitions through the ages from local to county matches have normally entered in couples but competed on an individual basis. In couples (ie. teams) they help each other with their equipment – it is a traditional thing. The excuse set out in your letter will make the British Government look silly – every fair-minded person will be disgusted. Furthermore, you are obviously discriminating against the entry into Britain of an innocent African-Rhodesian ploughman whose reward for his skill and endeavour you will now deny to him. Similarly, to support this lame excuse you threaten to deny entry to a Rhodesian ploughman who is a British Citizen and whose family home is in England.

This is really not good enough. We fail to see how by introducing politics into our World Ploughing Contest you can do any good towards solving the impasse between the British and Rhodesian Governments. This interference in our normal, friendly relationships between ordinary citizens of the world whose interest in life is to live and work together – who have only a desire to plough the same field together – is negative. If the British Government would show some goodwill and not prevent these two ploughmen from entering the country it would be a positive step towards better relationships and demonstrate that you do not discriminate against some and not against others.

These boys are not entering the country to demonstrate and be violent like some of those whom you freely admit. Furthermore, in previous correspondence both Mr Heath and Sir Alec Douglas Home have treated this matter with sympathy and, we thought support, especially when the previous government was in power. Also, on the assurance that persons with British passports or entry documents acceptable to the British Government would not be discriminated against and denied entry into the United Kingdom we embarked upon the promotion of the eighteenth World Ploughing Contest in Great Britain. As explained in my letter to Sir Alec Douglas Home dated 7th August, 1970 it was upon the strength of this assurance that the British Ploughing Association and the World Ploughing Organisation decided to hold the World Ploughing Contest here in England. Without feeling we had this assurance all plans would have been dropped. The British Ploughing Association also agreed not to host the Contest if the British Government applied discrimination against the ploughmen from any country.

At this late hour, after considerable funds have been expended you have let us down badly and a decision about cancellation has now to be taken in the light of whether or not we too have to dishonour our word as the British Government has done.

I ask you, please, to change your minds and appreciate that the ploughmen do not plough as a team. They plough for the honour of themselves and their fellow ploughmen. You can say the Rhodesian ploughmen no more represent Rhodesia than the British ploughmen represent the United Kingdom in the World Ploughing Contests. I hope sincerely you will with this explanation be more realistic and kindly and sensibly issue the necessary documents and allow Koti and Boswell to enter this country to plough. Meantime, we will not announce the decision set out in your letter of 21st June 1971.

Yours sincerely

Alfred Hall
World Ploughing Organisation

The Rhodesian Political Department of the Foreign and Commonwealth Office now scrutinised the World Ploughing Contest rules and pronounced:-

"... it seems to us that they clearly provide for countries to enter teams of not more than two competitors to represent them. There does not seem to be provision for individual competitors on any other basis."

We were asked to "kindly explain how it will be possible for two Rhodesian ploughmen to take part as individual competitors and not as a team representing Rhodesia". Such was the lack of knowledge in the Foreign and Commonwealth Office about ploughing! Had they never seen a ploughman ploughing all by himself? Doubtless, they might know how to swing an individual golf club. At some party they might even have sung "one man and his dog went to mow a meadow", but seemingly never raised their eyes from the golf links to espy one man and his tractor plough preparing the soil of a nearby field to grow food.

It became like conducting a correspondence course for the edification of the inmates of the Foreign Office on the fundamental, single-handed, oneness of the ploughman's labour. Lesson two had to be, more or less a repeat of the advice that the use of the word "team" is perhaps an unfortunate terminology since political ingenuity can apparently be used to imply an unintentional meaning. On the fifth of August 1971 the Foreign Office replied ". . . we do find ourselves in a genuine dilemma . . . " and again referred to the unfortunate letter which had gone from the member of the Ploughing Association of Rhodesia to the British Embassy in Pretoria. They pointed out that competition in the World Contest is "open to all countries in the world" and that the WPO affiliate in each country was "nationally representative". They cunningly failed to understand that WPO is an independent, international, non-governmental, non-political, nonprofit making, charitable organisation as are all the affiliated national ploughing match organisations in those democratic countries free of political dictatorship. The letter ended with an invitation: "if you feel that it would be helpful to come down to London to discuss the matter further, then we should of course be pleased to see you".

At the Foreign Office

I immediately telephoned the Chairman of the British Ploughing Association, Major Dick Gifford MC, at his farm in Wiltshire and asked him to meet me in London at noon on the morrow. Dick suggested we meet in Downing Street opposite the Foreign Office, then have a bar lunch whilst we talked the matter over before keeping our appointment. Next, I telephoned the Foreign Office, thanked them for their invitation and said that together with the Chairman of the BPA I would be there on the morrow at two o'clock in the afternoon. I had the impression they thought this was unexpectedly quick! The meeting time was agreed with the absence of Lord Lothian who was in Switzerland and the Head of the Rhodesia Political Department who had gone to Rhodesia.

At the appointed time Dick and I entered the stately Foreign Office building and were conducted to a small office to discuss with two representatives. The imposing architecture of the building belies it's inner appearance. We were not impressed with the drab furnishings and the need of some redecorating but, at least, we could feel satisfied that the authorities were not uneconomically lavish.

Dick Gifford and I emphasised that ploughing had nothing to do with politics and that ploughshares are the antithesis of swords. Further, that it must be clearly understood that every competitor in the Contest is a ploughman who has fancied his chance and on his own account entered a local ploughing match which he has won. The distant ambition of every ploughman who enters a ploughing competition is to eventually participate in the World Ploughing Contest.

Ordinary, down to earth people who have no political axe to grind use the terms "national", "nation", "country", "team" and "represent" in a perfectly natural and innocent way and have no thought that some political functionary will twist the original intended meaning of the words in a dishonest way. Politicians become dogmatic about terminology when they want it to have a particular meaning to aid their own political purposes.

Our discussions were courteous and direct. Dick and I were sharply critical when emphasising that discriminatory action was the mode of Nazis and Communists and should not be expected of the British Government. This was a shocking statement to make and they said so. Nevertheless, in the event of a humble ploughman being denied entry to the United Kingdom such Government action could be so interpreted and regarded on the other side of the world.

Having verbally reiterated all the facts in straight talking we sought the department's advice upon the opposites to all their objections. Their advice was that competitors must take part as individuals and not as representatives of their respective countries, that no national flags be flown, that the nameplates on Rhodesian competitors' tractors should show 'Ploughing Association of Rhodesia' rather than 'Rhodesia', and the list of competitors should be grouped under respective ploughing associations and not under countries.

Both Dick and I saw this as a breakthrough. The message had been received that ploughmen are individuals representing their worldwide brotherhood. In regard to national flags they do provide a colourful spectacle but few people have knowledge of which flag belongs to which country. Apart from national flags there are flags aplenty of other kinds including the flag of the World Ploughing Organisation.

We left the Foreign Office on friendly terms with the assurance that the outcome of our discussions would be presented to the Ministers for their decision.

On 3rd September I had a telephone call from the Foreign Office announcing the good news that the Ministers had agreed to our formula. This was confirmed in writing stating –

"arrangements are now in hand to restore United Kingdom passport and entry facilities to Mr Boswell and to grant a concessionary United Kingdom passport valid for one month to Mr Koti".

Our reply was

'Dear Lord Lothian,

We are well satisfied with the sensible and proper decision you have taken to permit the entry into this country of the ploughmen from Rhodesia so that they can participate in the eighteenth World Ploughing Contest. All arrangements are made according to the formula we agreed and everyone is content.

We thank Sir Alec Douglas Home and yourself for your sympathetic understanding.'

Meanwhile, Lord Lothian kindly informed, and pleased, Willie Whitelaw MP (late Viscount Whitelaw), Lord Craythorne, President of the British Ploughing Association and all others who had made representations to help overcome the problem. The sequel was described in an appreciative letter from Richard Boswell who first flew to South Africa where his passport was validated at the British Embassy in Pretoria and upon arrival in England wrote,

'Dear Mr Hall,

Just a note to let you know that we arrived without any passport troubles at all ... Thank you very much indeed for all the trouble you have taken in helping us.

Richard Boswell.'

During all this time of controversy, correspondence, debate and negotiation the matter was never revealed to the press. Both the British and Rhodesian Governments, the WPO, the BPA and the Ploughing Association of Rhodesia kept the affair under wraps. The Rhodesian press began to sniff but Eric Linnell managed to keep the problem confidential. One of the press agencies was a little inquisitive but, fortunately, there appeared no sensational or provocative newspaper stories which could encourage the political pushers to pester the ploughmen's platform.

Of the forty-two ploughmen from twenty-one countries, Danish farm manager Peter Overgaard won the world title and the Runner-up was New Zealand farmer Alan Wallace.

Chapter 17

THE CONTRARY AIM of COMMUNISM

Between the years 1965 and 1971 WPO was embarrassed by the international political aspirations of the East Germany Communist Government. The ploughing organisation in the German Democratic Republic was part of the Ministry of Agriculture that provided the sum of about one million East German marks annually for the promotion of ploughing matches and to cover the cost of sending the DDR ploughmen to the World Ploughing Contest. Therefore, the DDR placed great importance upon the use of the initials "DDR" or "GDR" and the flying of the DDR flag.

This was an entirely different situation from that of the Ploughing Association of Rhodesia and also of the British and other independent ploughing match societies in democratic countries of the west. The ploughing committee of the DDR Ministry requested agreement on three essential points

"1. That the official designation of the East German state is 'German Democratic Republic' (abbreviated GDR) at all World Championships, in official documents and on all other occasions where the Ploughing Committee of the GDR is mentioned.

2. That the flag of the state of the German Democratic Republic be hoisted at all World Ploughing Contests at which GDR ploughmen take part and where flags are hoisted. Likewise, appropriate observation to be accorded the national anthem and emblem of the GDR.

3. Before the World Ploughing Contest is allotted to a country the organiser should warrant that this decision will be fulfilled."

The ploughmen from the DDR would be unable to participate in the World Contest if these requirements were not met.

This was clearly a case of the ploughmen officially representing the government of their country. Whether or not the individual ploughmen did so in their hearts was doubtful. History shows they were pawns in the political game when the eastern part of Germany was not recognised as a separate country. There was no western diplomatic representation in the so-called DDR.

In West Germany the cry was "Germany is indivisible". As a token of this indivisibility people wore a small lapel badge replica of the Brandenburg Tor – the famous centrepiece of Berlin, closed by the Communists and flanked with the massively obscene east-west dividing wall. In the great exhibition halls of the German Federal Republic were displayed the flags of the provinces annexed into the east zone "As a tribute to and reminder of our abducted relatives and friends on the other side of the fence". People could not come to terms with the idea that Germany be divided into two separate countries. There was no East German DDR so far as the west was concerned.

The WPO Governing Board agreed in principal with the DDR requirements so far as they were in keeping with the policy of the International Olympic Games Authority whom we had consulted.

The WPO also declared that the same requirements, rights, privileges and other conditions applicable to the DDR representation would be applicable to the representation of all other WPO affiliates, including Rhodesia, at all World Ploughing Contests. This would be the policy of WPO notwithstanding the possibility of any host country government imposing political restrictions over which the WPO had no control.

The situation now was, simply, that unless WPO could guarantee to fly the DDR flag at World Contests East German ploughmen would be withdrawn in protest. However, WPO had no exclusive right to fly the flags of other lands in any country. We had previously been advised that when the World Contest was to be held in Norway in the year 1965 there was no possibility to fly the DDR flag. The DDR was not recognised as a country, nor was the DDR government recognised. Obviously the authority to fly national flags is not the prerogative of the World Ploughing Organisation, or of any other organisation, but is determined solely by the government of the host country. The same ruling would apply in New Zealand (1967) and Denmark (1970).

In response to an invitation to attend the DDR Ploughing Championships, making travel arrangements to visit a country not supposed to exist was complicated. However, with the help of the ploughing fraternity in the DDR the necessary visa and travel documents came to hand. The visit gave the opportunity to observe an entirely different and imposed lifestyle. The ploughmen were the same friendly fellows like one meets elsewhere in the ploughing world. The match was on a big scale, well organised and attended by a large group of government officials of whom, I understood, three were Ministers of Agriculture of varying degree. From a long wide stage suitably decorated with Communist party emblems and DDR flags the Ministers made long speeches extolling the system, the ploughmen and the collective farms they had the privilege to plough, but condemning the dangerous people on the other side in the west. Included on this loaded stage midst Communist government and party men sporting open-neck shirts, stood their guest from the West wearing a "Golden Plough" necktie and, consequently, feeling distinctly unlike a member of the proletariat.

On the serious subject of displaying national flags there was no compromise. The flag of the DDR must be flown with all the other national flags at the World Contest. They put the responsibility on the WPO and did not want to hear differently. Their political desire could be understood but their determination to ignore the fact that non-communist countries would not allow their flag to be flown was pointless. It was better to talk more about the practical side of ploughing. They accepted the suggestion that without flags all nationalities would be treated the same, and were told that WPO could meet the situation by flying only the host country flag in company with the flag of WPO. WPO was anxious that East German ploughmen should have the chance to participate with their world-wide brother ploughmen.

My assignment at the Federal Ministry of Agriculture in Bonn met with a similar determination to that experienced on the east side of the Iron Curtain. The West German Government did not recognise the East German Government. So far as the West was concerned the DDR was not a country.

"What about the ploughmen in the east?"
"They belong to greater Germany and we welcome them as our own German people."

It was an odd discussion. Surely, it would be making progress to bring them together in the World Ploughing Contest? As for flags, there was one German national flag for all Germans. Again, one could understand their point of view and their desires, and also recognise that expressed political desires can be very different from the underlying peaceful desires of ordinary folk to live free from political partition.

Next, on the diplomatic circuit, was a visit to Norway to consult at the Ministry of Foreign Affairs in Oslo where there was said to be information about some kind of so-called "Oslo Formula" about flying of international flags. The detail was so unclear that it seemed less reliable than to fly no flags at all. The Foreign Office Official stated categorically that if there was a so-called DDR flag it could not be flown in Norway. The man at the Ministry agreed that to fly only the flag of the host country and the flag of WPO would be the most satisfactory, even though maybe a little disappointing for those ploughmen whose national flags were legitimate. The display could be enhanced with the flags of local towns and counties if desired.

Thus it was, and after all the work behind the scenes the two East German ploughmen who happily competed in the 13th World Ploughing Contest, at Ringerike in Norway, in 1965, were thirty-four year old Alfred Lehmann and twenty-one year old Wilfried Domke, – without flags. Ploughmen from the DDR were entered in no more than three World Contests but participated in only two. In the year 1967, Alfred Lehmann and Klaus Zimmerman were withdrawn from the fourteenth Contest in New Zealand. Finally, Klaus Binder and Gottfried Prohl competed in the sixteenth Contest in Yugoslavia in 1969 – with flags.

At the WPO Board Meeting in year 1970 the ploughing committee of the DDR proposed in writing that four-furrow ploughs might be used in the World Contest instead of two-furrow ploughs because the ploughmen in the DDR were accustomed to four-furrow ploughs and were unable to participate in the World Contest under present rules for two-bottom ploughs. The WPO Governing Board decided that the idea be referred to all the affiliates for their comments to be considered at the next meeting of the Board. At that meeting in year 1971 it was decided "that competitors shall, at their own discretion, use either two-furrow or three-furrow ploughs in the World Ploughing Contest".

In the year 1972 the ploughing committee of the German Democratic Republic withdrew from membership of the World Ploughing Organisation with the accusation that it was "too political"!

Chapter 18

NO POLITICS, BUT
"A MAN WHO TELLS JOKES"

Thankfully the year 1972 brought a brief relief from the political parleying when the 19th World Contest was held at Vernon Centre in Minnesota, USA. Firstly, and regrettably, the East Germany Ministry of Agriculture's "Komite für Leistungspflügen beim Landwirtschaft der DDR" had defected from WPO preferring political doctrine for their ploughmen to be more important than the advantages of international fellowship.

Secondly, there was no objection against the entry of Rhodesian ploughmen Sydney Moorcroft and Richard Boswell into the United States of America as competitors in the World Contest. Champion ploughmen from twenty countries participated, including two from Kenya and one from Hungary. The venue was the fertile district known as "Blue Earth Country" on the seven hundred acre Shadybrook farm of cowboy-hatted, flying farmer Bert Hanson. The Contest was supported by an Exposition portraying a countrywide variety of aspects of American farming, manufacturing, culture and, particularly, entertainment. The event was designated an International Trade Show by the USA Treasury and Department of Commerce thus authorising foreign exhibitors to bring their products for display without having to pay Federal import tariff. There were no political problems but there was an anxiety of a different kind.

As might be expected, there was an imposing display of American showmanship; a profession in which they are very adept. Showmanship can be simple or full of flair and, sometimes, without meaningful ceremonial. This became a source of irritation when we tried to confirm the pre-determined ceremonial arrangements for the unveiling and dedication of the commemorative Cairn of Peace.

These monuments dating since the first World Ploughing Contest held in year 1953 at Cobourg, Ontario, in Canada are spread around the world, one at each place where a WPO Contest was held, and represent links in a chain of fellowship. The first piece of ceremonial is when the ploughmen parade their tractors and ploughs from the

Marshalling Yard through the exhibition area to the competition plots. The centre piece of the Exhibition at Shadybrook Farm was a high and wide covered stage from which one of America's best known farm radio broadcasters Herb Plambeck, who founded the American ploughing championships, read a commentary on each ploughman and his equipment as he drove along. During the rest of the day the stage was occupied variously by the US Air Force Band, and entertainers from film, television and theatre.

The General Office for the whole event was that of the Co-ordinator on whom we relied to implement WPO's specific needs, as per contract. Upon calling to check the final detail for the Closing Ceremony – unveiling Cairn, Dedication, Lowering of Flags – he and his staff were in lively discussion. In order to review the procedure with the principal players in the ceremony it was necessary to talk with the special person who was to perform the unveiling, the Church dignitary who would declare the Dedication and the Conductor of the Band which would play during the Lowering of Flags. All involved had to be acquainted with the timing and the signals or cues.

The office was full of people seeking a variety of exhibition ground information, but the staff was pre-occupied with the "Bugs Bunny Show" and the ninety minutes that Bob Hope was to hold the stage with the Warren Bills Orchestra! Herb Plambeck and the WPO General Secretary were both wishing to get a word in edgeways. This must have been galling for Herb whose loquacity was unaccustomed to restriction. It was time to adopt an artifice that would call attention to ourselves. After all, _we_ were here for the ploughing and but for the ploughing the others would not be here.

With casual innocence and uninhibited audibility I turned to my friend "Herb! Who _is_ Bob Hope?". The room went immediately silent. All eyes turned to look for an obvious ignoramus. Herb's passive reply "Well, he's a man who tells jokes". Whilst the rest stood agog, and before they could recover, we stepped forward to command the Co-ordinator's attention to check the arrangements for the Cairn ceremony due in three hours time. Were the principals available for a simple rehearsal?

To our dismay they were not. The stage was ready for Bugs Bunny followed by Bob Hope. The monument was ready for WPO, but the appointment of a VIP to perform the unveiling, a Churchman to declare the Dedication, and a Band to play the music had been overlooked!

In the course of two hours before the time for the Ceremony the Co-ordinator's helpers found a willing Clergyman among the spectators and a van fitted with loudspeakers, a gramophone and music discs to play the USA National Anthem. Since no American VIP was available so late in the day, the duty of unveiling the Cairn fell to WPO Chairman Walter Feuerlein who made an appropriate short address in his naturally good-humoured manner.

This ceremony was in sharp contrast from that of the 6th World Contest at Peebles, Ohio in year 1957 when a member of the USA Government spoke and was assisted by a charming young "Queen of the Plough" whilst the US Air Force No 1 Band played during Lowering of Flags.

Apart from that one lapse of organisation the Mankato Contest was very successful. The winner was Willi Flatznitzer from Austria and Mads Bakken from Norway was the Runner-up. The post Contest hosting was enjoyed in family homes and communal entertainment and festivities, establishing yet more friendships across the world.

Chapter 19

POLITICAL INTERFERENCE

For the next few years from 1973 World Contests were given headlines in the press which we did not desire. Sensation-seeking reporters overlooked the peaceful purpose of the ploughmen and the proficiency of their ploughing. Contentious aspects of politics were introduced by persons with no interest in ploughing but for whom an international event could provide a fortuitous platform.

Some competitors arrive for the Contest several days earlier than the official arrival date to do some private practice by arrangement with local farmers. Several did so in advance of the twentieth Contest in year 1973 at Wexford in the Republic of Ireland. Among those who did were two ploughmen from Kenya and two from Rhodesia. The Irish public relations officer arranged an early press Conference in Dublin with the newly arrived ploughmen. After a wealth of information had been given about the Ploughing Contest and associated events one reporter wanted to know how was it possible for two Kenyans to sit on the same platform with two Rhodesians? He would like the ploughmen to answer. Their reply was to the effect "Why not? We are all ploughmen; we do the same work, we are friends, we come from Africa and we have qualified to compete in the World Ploughing".

After the Conference the four ploughmen walked the streets together shop window gazing. A photograph of them appeared in the 'Irish Independent' on the 27th September 1973. This fraternising plus the acceptance of Rhodesian competitors in the Contest and into Ireland aroused the ire of a small group of 'stirrers' in Dublin, not all Irishmen, who provided Irish newspapers with letters and interviews. They must also have made representations to the governments of Kenya, Yugoslavia and Czechoslovakia. The two ploughmen from Kenya were Joseph Kangogo and Kuru Njungwa accompanied by their coach Joseph Gichungwa. Joseph received a telegram via the Kenyan Ambassador in London with an order from Nairobi that the Kenya ploughmen must not be accommodated under the same roof as the Rhodesian ploughmen, withdrawn from the Contest and return home.

WPO headquarters was Whites Hotel, in the county town of Wexford, where the WPO General Secretary's office was upstairs completely shut off from the public part of the hotel. Behind locked doors Contest Co-ordinator Dermot Jordan and the General Secretary interviewed the Kenya lads. They were nervous and sad. Their great chance of a lifetime was to be snatched away. They left us in no doubt they did not want to be withdrawn. The press had become alert to the fact that a situation had developed. Reporters waited impatiently outside our locked office door.

The Kenya and the Rhodesian ploughmen had no quarrel between themselves. It was important to avoid any report going back to Nairobi to say they were living under the same roof and in contact with each other. It was important to handle the affair in strict confidence not only to avoid unwelcome publicity, but in the interests of maintaining morale among the other ploughmen. The less they knew the better. They were all here to plough and to enjoy each other's fellowship.

To cope with the situation we smuggled the three Kenyans by night through the back yard of the hotel and drove several miles to a small but good restaurant, had dinner together instead of dining with the group at headquarters; then accommodated the ploughmen with a very friendly Irish farming family.

Joseph Gichungwa was driven to Dublin airport and flew to London to appeal to the Kenyan Ambassador. He returned next evening disappointed. The Ambassador had gone away and the second man was not at all helpful. Next I spoke on the telephone with the authorities in Nairobi and explained the impossibility of returning the ploughmen to Kenya. To do so would offend the Irish people who had spared neither themselves nor expense to provide the warmest of welcomes and hospitality. Furthermore, there was the problem of booking and paying for return flights on a change of date. Agreement came that the Kenya ploughmen could stay until the day of their flight reservations but must not appear on the Contest field or with the WPO group. Suffice to say that despite their bitter disappointment everything was done to give them a pleasant holiday.

Midnight knock

On the night following there came a midnight knock on my bedroom door. My callers were the coaches for the Czech and Yugoslav ploughmen. I was shown an urgent message from the Czech Trade Mission in Dublin which read "Wife of Mr Abosi is seriously ill. She is

118

also sister of Mr Milik (the two Czech ploughmen). They are asked to return home." In great distress the Czech coach apologised and said they would leave now, during the night, and go to London where the Czech office would see them off to Prague. Would I please see that their ploughs and tractors were returned to the lending companies and express their deep regrets and sincere thanks? Likewise to please express apologies to their Irish hosts and to the WPO Board and convey grateful thanks for all kindness they had received. They desired to be away before morning rather than have to face everyone and say an early goodbye. The Yugoslav coach said he had not yet received such an order from his government but felt sure that one would come. He wanted me to know that he expected the same problem.

Both said neither they nor their ploughmen wished to leave. When I suggested they might take no notice they remarked that was not possible under the system of their regime. They appreciated it would be an act of bad manners to go and of which they would feel ashamed. Anyhow, the hour was too late to set out for Dublin so the Czechs stayed overnight and quietly left the next day.

In the meantime no official order came to the Yugoslav ploughmen. They continued to practice and competed in the stubble ploughing on the first day. Then, that same evening they did receive an order not to participate in the Contest. To their utter disappointment Stefan Bukwic and Marjan Frelic were obliged to withdraw from ploughing the grassland on the second day of the Contest.

Midst all this bother the fellowship among the international ploughmen and their respect for each other was in no way impaired. None of them wished to withdraw, nor did any of them wish any other to withdraw. This grass-root people to people compatibility does not seem to be understood by minds saturated with political bigotry.

During the pre-Contest preparation and practice days the event was a daily sensation news item with bold headlines in all the Irish newspapers. It was not the peaceful endeavour of skilled ploughmen promoting a better understanding of improved soil tillage that was of major interest. In some reports one could substitute the word 'politics' for the word 'ploughing'. When the ploughmen were asked to comment they did not enter into controversy. The Rhodesian coach, Geoff Oliver, said "As far as we are concerned there is no controversy. We are here to compete in the ploughing and we are not interested in the political aspects of the situation".

The Kenyan coach, Joe Gichungwa, said "Ploughing and politics are totally different matters and we want them kept apart".

The small group of objectors in Dublin must have worked very hard to give the World Ploughing Contest so much publicity in their pursuit of propaganda for their own political purpose. Unfortunately, their activity embarrassed the Irish Government who would have been happy to bask in the reflections from a truly rural international ploughing match and associated formalities, religious ceremony, banquet and festivities in the Irish tradition of "a thousand welcomes". Instead of the nation's leaders being officially involved in all the functions as was intended they were demure and 'sat on the fence' in a pose of neutrality.

Demonstrators

The objectors whose leader was of foreign nationality grew more vociferous. There was a hint of possible violence as a means of preventing the Rhodesians from ploughing their plots. An advertisement appeared announcing that two coaches would carry objectors from Dublin to the Contest at Wexford to demonstrate.

Michael Connolly, the Irish Board Member of WPO, advised the press and any objectors that the aims of WPO are to 'breakdown political barriers and promote fellowship and peace'. That this has been an achievement of WPO is patently clear since overcoming the Irish problem of year 1954. Nowadays no two ploughing organisations support each other's events better than do the National Ploughing Association of Ireland in the Republic and the Northern Ireland Ploughing Association in the six county northern province.

Michael Connolly announced that the ploughing venue would be attended by a large number of stewards who had no intention of becoming involved in violence. The demonstrators had the right to picket peacefully but they would be dissuaded from interfering with the competition. The announced attendance of busloads of demonstrators was, of course, of interest to the police. A senior officer in plain clothes came to WPO headquarters at Whites Hotel. He assured me that in the event of trouble his force was ready to deal with any situation. He also said that one of his plain-clothes garda officers would become a member of our WPO group and live in the hotel. The garda officer told me he was not too concerned about possible interference with the ploughmen but more about in what trouble the demonstrators might find themselves.

On the eve of the Contest an announcement appeared in the Dublin newspapers "Ploughing Protest Called Off" by the anti-apartheid movement since "no useful purpose would now be served by a protest at the venue itself" . . . and "we do not wish to mar the enjoyment of thousands of farmers and their families who have been looking forward to this event".

Bearing in mind the Police Chief's intuition it became clear they did avoid trouble by absenting themselves, except for one man. The pouring rain was driven horizontally by Atlantic gale force winds and the feet of thirty thousand visitors churned the rich soil into ankle deep, slimy, runny mud. Both days the event was an outstanding success despite the inclement weather. It was not uncommon to see women who had arrived without rubber boots carrying their shoes and stockings as they plodged barefoot through the mud.

At the Festival Evening in the Wexford town theatre the Golden Plough was presented to new World Champion Paavo Tuominen, from Finland, and the Silver Rosebowl to Ireland's John Treacy, the Runner-up. Afterwards a group of Irish farmers invited me to accompany them to the bar. Here they enthused upon the success of the event with one expression of regret. I was curious. With a particular eagerness they had looked forward to the arrival of the demonstrators. Their non-arrival was a disappointment. I was more curious. They explained.

This group of hardy Irish farmers, one of whom had been a Commando Major in the British Army, were the nucleus of a larger group who had followed events with absorbing interest and deployed themselves ready for what action might be necessary to frustrate any interference with the competitors. They said that in the event of violence their plan was to deposit the demonstrators in the nearby river.

Much as I appreciated their thoughtful protection and understood their disappointment I felt thankful they missed out on that particular bit of fun. At the back of my mind I knew for whose safety the garda chief had been concerned.

A confession

The sequel to this episode was told in a confession made to me several months later by sturdy Welsh farmer Jack Tanner. Jack was a plot steward and supervised a corner plot ploughed by one of the Rhodesian competitors. Along came the only dissentient voice among the thousands of spectators. He was a loud and vulgar mouthed fellow who shouted obscene abuse at the Rhodesian. There were ladies about when Jack dutifully approached the chap telling him to curb his language and not interfere. Whereupon, the demonstrator threatened his violence upon Jack!

Jack Tanner was short, stockily built with broad shoulders and strong arms built as for Cumberland and Westmorland style wrestling with "wood under his oxter". Fortunately, the ladies and other spectators plodding through the mud in that corner of the storm battered field had passed out of earshot and sight when Jack's horny fist shot out and knocked the intruder full length into the mire. Not only did he fall but slid full length in the puddle from whence he gathered himself and departed in great discomfort, threatening he would be back to 'get' Jack.

Jack revealed his story to me in response to a chance conversation. We were talking about the World Contest and referred to the Festival Evening at which Mr and Mrs Francis Leigh, on whose farm the Contest had been hosted, were to their surprise brought on stage to receive a presentation. Jack Tanner casually admitted he was not there. As soon as the Contest ended he returned to his hotel, packed his bag, boarded the night boat at Rosslare and sailed home to Wales. Why? He was so alarmed that his antagonist might keep his threat and arrive with a gang to seek him out. Jack did not want to be the reason for a disturbance at the Festival Evening, so he decided that discretion was the better part of valour and did a flit.

He need not have worried, the theatre was crowded with good humoured, hospitable folk and a most enjoyable evening was had by all.

Chapter 20

A THREAT OF ASSASSINATION

Whilst politicians squabbled over Rhodesia the governments of Finland, Sweden and Netherlands would not grant entry permits to Rhodesian ploughmen to participate in the World Contests held, respectively, in years 1974, 1976 and 1977. Nevertheless, the scheduled plan for future Contests went ahead. Everything seemed set fair for freedom from controversy in Canada in September 1975, but no sooner had everyone arrived when, sadly, the two Kenya ploughmen, Kuru Njuguna and John Imithira, received orders from Nairobi that they must return home. This was bitterly disappointing for Kuru who had qualified on two occasions to participate in the World Ploughing Contest and was twice forced to withdraw. Thus the ploughmen were denied of their lifetime opportunity not only to plough in a prestigious Contest but, also, from experiencing life and hospitality in Canada, and sharing fellowship with ploughmen from other lands.

Once again the reason for the withdrawal was declared to be because the champion ploughmen from Rhodesia were also participating. Again the political aspect made headlines in the newspapers. Thankfully none of the other competitors suffered the ills of political pollution.

This twenty-second Contest was held at Oshawa in Ontario where the Opening Ceremony was to be performed by Prime Minister Pierre Trudeau. He would make a speech and unveil the Cairn of Peace, built in the shape of a large Maple Leaf inset with the traditional inscribed blocks of natural stone brought from the ploughmen's' homelands. The ceremony was timed for one o'clock in the afternoon and the Canadian Broadcasting Corporation had arranged to broadcast the proceedings live over the coast to coast network. The Prime Minister would arrive on the ground shortly before the appointed time. All other civic and parliamentary dignitaries forgathered for lunch in a large marquee. I, as General Secretary, was to make the introductions from the high podium beside the monument, and the CBC radio producer emphasised the importance of making a prompt start at one o'clock, otherwise during delay nothing would be going out on the broadcast.

At lunch table I sat with former Ontario Agriculture Minister William Stewart and Chief Police Officer Ralph Coleman. Next to Mr Stewart was the local Member of Parliament and elsewhere an array of Mayors and other dignitaries and civil servants. Lunch finished neatly in time for the expected arrival of Mr. Trudea when, at ten minutes before one o'clock, a messenger arrived to whisper into the ears of the Police Chief and the local Member of Parliament Norman Cafik. Prime Minister Trudeau had flown from Ottawa to Toronto airport and upon arrival there decided to travel no further. His appearance and his speech at the Ploughing Contest were cancelled. It was said he made his decision just before touching down at Toronto International Airport en route to the Contest, whilst being briefed by his aides who referred to the list of ploughmen which included two from Rhodesia.

What to do now? There was a crowd of several thousand people waiting for the ceremony and the CBC was ready to go 'on air'. My immediate reaction was to turn to Mr Stewart, whom I knew from our earlier Contest in Canada, and say "Bill, I must ask you to welcome the ploughmen and all the people and you to unveil the Cairn". Mr Stewart hesitated. There were only a few minutes left to go and mount the platform. I pleaded "It would be a discourtesy to all the international guests and to all the spectators if no one would so much as say a Canadian welcome and declare the event open".

"You are right. I will do it" agreed William Stewart, and we walked together sharp but with composure onto the stage and up to the microphones. The CBC producer signalled with a wave of the hand that the coast to coast broadcast was on. As I stepped up to the lectern and before I could speak there was a demonstration orchestrated from every direction in the large crowd. Slogans and crisp messages were bawled from a loud hailer and placards hoisted all around. Then in a matter of minutes the placards disappeared, the shouts ceased and all was quiet again. I gave the announcements including the last minute news of the absence of the Prime Minister. Bill Stewart made a warm speech of welcome and Mr Cafik MP explained that "because of international complications" Mr Trudeau could not appear. As reported in the press Mr Cafik "hastened to add that the decision did not reflect on the Prime Minister's affection for farmers".

The unexpected, vociferous demonstration had nothing whatever to do with the participation of the ploughmen from Rhodesia. The demonstrators fully expected, as did everyone else, that Pierre Trudeau was among the platform party and ready to step up to the battery of microphones. The placards and loud hailing were a protest

124

against the expropriation of farmland for a proposed new airport at nearby Pickering. Everyone got the well-staged message both on the field and right across Canada by radio except Mr Trudeau at whom it was principally aimed. That the Prime Minister should have abandoned his journey at Toronto airport upon the excuse that he only then learned of the ploughmen from Rhodesia was very odd. One Canadian described it as "just a lot of pure, cruddy bumbling". Perhaps this description was more than apt since one MP spokesman said in an interview that he learned some weeks in advance about the entry of the ploughing champion and his runner-up from Rhodesia and understood the Department of External Affairs and the Contest organisers had straightened the matter out. For the record, there had been no discussion on the subject with either the Canadian host organisation or with WPO.

There was a more sinister aspect which was suggested might have persuaded the Prime Minister not to fulfil his engagement. It was reputed to be a threat of assassination. The rumour was first whispered and then reported in the press. The 'Toronto Star' reported on the 24 September 1975, 'Royal Canadian Mounted Police guarding Prime Minister Pierre Trudeau were alerted yesterday (three days before the first day of the World Contest) after two men were overheard talking at Toronto International Airport about killing Trudeau. A Ministry of Transport employee at the airport told the RCMP one of the men said something like "now is a good time to kill Turdeau". Chief Superintendent Don Wardrope said the police considered the threat "vague and veiled", but the Royal Canadian Mounted Police in Oshawa were told to be extra careful'.

As soon as the demonstrators melted away in the crowd the ceremonial programme was successfully accomplished. The Contest was praiseworthily described as efficiently organised and enjoyed by an attendance of one hundred and sixty thousand spectators. Twenty-nine year old Gunnar Hersleth from Norway won the "Golden Plough" and Rund Hermus of the same age from the Netherlands and twenty-six years old Carsten Balle from Denmark tied for second place.

Chapter 21

A VARIETY OF INTERESTS

As already mentioned the champion ploughmen from Rhodesia were denied entry to Finland, Sweden and the Netherlands. Controversy continued in the press of each country but their absence provided no scope for sensational stories.

Advance planning visits involve meeting a variety of people in the cause of public relations, in company with the host country WPO Board Member, for appointments, committee meetings, interviews and press conferences. On arrival in Finland one can expect an early expression of hospitality will be an invitation to experience the sauna – to alternately stew, cool, stew and sweat in the hot wood-lined cell; then after plunging into a cold swimming pool, relax with a drink, bread and cheese and a long, peaceful rest in an armchair.

WPO Board Member Walter Breitenstein had his log fire burning merrily and his sauna temperature at peak. The sauna was Walter's recipe for the revival of weary travellers in readiness for subsequent negotiations. In this connection the effect of the Finnish sauna is probably more restorative than basking in the sunshine on the hot sand of a Mediterranean beach. Thus cleansed and refreshed one replaces the weight lost during perspiration with a hearty meal, whilst being briefed about the programme arranged for one's visit. Finns are firmly of opinion that a most favourable venue for the conduct of constructive negotiation is a good, spacious sauna in which the negotiators are as natural as it is possible to appear and flamboyancy has no part. There were at least three more sauna bath meetings in the itinerary. My visit coincided with a high-priority USA-USSR "Cold War" meeting in Helsinki and the government authorities had, apparently, gleefully reserved all the public and commercial saunas for the relaxation and refreshment of their east-west guests on the precept that they could, at least, be in agreement in the sauna!

The host committee meeting for the forthcoming World Ploughing Contest was held in the sauna at the Helsinki offices of the Finnish Association of Agriculture and was attended by the Minister of Agriculture Heimo Linna. There we all sat, sweating profusely in our birthday suits discussing the programme for the twenty-first World

Contest. From time to time we trotted out to "enjoy" the shock of a cold shower. Unadorned, we could only talk, not write notes whilst stewing. Notes were written up afterwards in the recovery lounge between snacks of bread and cheese washed down with weak beer. This was the only time I met a Minister of Agriculture in the pristine state of nature.

Of the other meetings the best sauna of all is worth recalling. It was on a small island in the Baltic Sea, a rather choppy sailboat distance from the mainland. The log fire soon brought the inside of the thick-walled timber hut to oven temperature and courage was demanded to run along a plank against a strong, cold wind and jump into the uneasy sea. But the feeling of rejuvenation after swimming back to shore and entering the shelter and warmth of the sauna hut was well worth the effort.

Some eighteen months later, Minister of Agriculture Heimo Linna hosted Finland's Welcome Reception to the international ploughmen upon arrival at Helsinki for the 21st Contest in the series. Mr Linna greeted and shook hands with his guests as they entered the great banqueting hall. He was a farmer himself, a small man, in an immaculate suit as I had not seen him before, and so would probably not have recognised him on the street. Upon exchange of greetings and recalling the earlier ploughing meeting in the sauna, it was amusing to quip this was the first time we had met with clothes on!

The twenty-third and twenty-fourth World Contests held on the Bjertorp Estate near Vara in Sweden (1976) and at Flevohof in the Netherlands (1977), respectively, both enjoyed the patronage of each country's Royal family. In Sweden, farmer King Karl XVI Gustav declared the Contest open, unveiled the commemorative cairn and greeted each individual ploughman with a welcome on behalf of the Swedish nation. Likewise, in the Netherlands Queen Beatrix and her husband Prince Klaus performed the ceremonial and paid a personal tribute to each ploughman.

A distinction between the two Contests was that in Sweden the event was supported by a large exhibition of agricultural machinery and products whilst in the Netherlands the farm machinery manufacturers and dealers maintained a different policy. They strictly limited themselves to exhibiting at only three major annual Shows in the country. Not even a world event could persuade them to do otherwise. The absence of an attractive commercial exhibition did not make the World Ploughing Contest held on the Polders less popular with forty thousand spectators.

Instead of pursuing the commercial aspect the Dutch are adept at promoting a cultural exhibition of rural art and crafts, displays of vintage tools, young farmer activities, animal husbandry and general aspects of country life. At the World Contest in the year 1977 the WPO affiliates were invited to make an international craft display with traditional specimens.

It is appropriate that the traditional aspects of rural life should be associated with, and so far as possible displayed at a ploughing match. It is interesting to know that the world's ploughmen are not dedicated to ploughing as their work and hobby to the exclusion of other interests. The biographical notes which regularly appear in the WPO Contest handbook reveal involvement in a wide variety of interests and activities. These include music, swimming, acting, pottery, horsemanship, dancing, motor sports, field sports, flying, reading, country crafts, athletics, and among many others the care and maintenance of both vintage and modern machinery.

Since the ploughmen are from widely diversified landscapes and climates – from the mountainous regions of Scandinavia to the flat land of Holland; from the great plains of North America to the rolling farmland of Australia, and many places in between, their activities illustrate the natural facilities of their respective environments. For example, where snow lies deep on flat land then snow-mobiling takes over from ploughing; where there is just as much snow but the land is hilly the ploughmen go skiing. In general, from wherever they come they share a common interest in many things, be it mechanics or music, crafts or camping, Church and community. It has been said with truth that good ploughmen make good citizens.

The two champion ploughmen from Rhodesia were not the only absentees from both the Swedish and Netherlands Contests. For entirely different reasons so were the ploughmen from Kenya, Yugoslavia and Hungary. The Agricultural Society of Kenya of which the Kenya Ploughing Organisation is a part announced their decision to temporarily withdraw from membership of the World Ploughing Organisation. Membership was resumed in the year 1981 for the twenty-eighth Contest when the sole competitor from Kenya was Isaac Kimongo at Wexford in the Republic of Ireland. The absence of competitors from Yugoslavia and Hungary was for financial reasons. From time to time, WPO affiliates in eastern bloc countries failed to raise funds enough to reward their champions with a travel prize to the Contest. This is somewhat singular when one considers that no expense seems to have been spared to ensure participation of their athletes and gymnasts in international competitions.

Chapter 22
EPISODE IN NORTHERN IRELAND

When the 7th Contest was held at Armoy in Northern Ireland in year 1959 there were no political problems that affected WPO. Although, there was some conflict between some oil companies as to whose product serviced which competitor! By the 1970s decade political violence was rife in the province and Northern Ireland had a very bad press worldwide. The Northern Ireland Ploughing Association invited WPO to stage the 23rd Contest there in year 1976, but in view of the killings the WPO Board Members were reluctant to return to Northern Ireland. The majority of them had been told by their respective national organisations not to accept the invitation and that, in any case, their ploughmen would not be entered.

This was a big disappointment for WPO Board Member Tom Reid who insisted there would be no danger to anyone attending the World Ploughing Contest. He maintained that many news media reports were exaggerated. The WPO Board's careful and sympathetic discussions were considerate of two aspects. The first was the question as to whether or not the competitors for the year 1976, as yet unknown, would themselves agree to travel to Northern Ireland; might they be dissuaded by wives and families in view of what they read in the newspapers? Secondly, WPO did not wish to let down the Northern Ireland Ploughing Association. If there should be any doubt then it was right and proper that the NIPA should know the position before committing their organisation to great expense and probably loss of funds.

To properly assess both support and opposition the Board decided the Chairman, General Secretary and Treasurer pay a fact finding visit. Chairman Arie Stehouwer from the Netherlands, Vice-Chairman Franz Geiger from Austria and myself made the journey. Our terms of reference were to faithfully and factually report our findings and impressions to each member of the WPO Board for reference to each respective affiliate. After studying the report each affiliate would then return their answer to a questionnaire. If half the answers were negative then the Contest would be switched to Sweden.

During our stay in Northern Ireland we had discussions with the

Chairman, Vice-Chairman and Secretary of the NIPA, the Deputy Chief of the Northern Ireland Police Force, senior civil servants from the Department of Agriculture, the Principal of Greenmount Agricultural College and the owner of the land offered for the Contest. Our subsequent report was long and detailed. We travelled by sea and were surprised to find entry and security formalities less than expected. We learned that other events of an international character were held successfully in Northern Ireland and were assured the police would gladly provide their usual service the same as in any other country. Greenmount Agricultural College was available as Contest headquarters for accommodation and the site alongside the coast of Lough Foyle was a polder some three metres below sea level protected from the sea by a high dyke.

These and many other aspects involved in planning a World Ploughing Contest from finance to food were all taken into consideration. Our conclusions were summed up in our unanimous recommendation "that from the information supplied to us and from our personal observations, and subject to the provisions we have herein explained, it is reasonable to proceed with arrangements to hold the World Ploughing Contest in Northern Ireland in the year 1976".

At the annual meeting of the Deutscher Pflugerrat (German Ploughing Society) to launch plans for the twenty-fifth Silver Jubilee Contest the members posed questions about Northern Ireland. The most direct came from a father who asked, "Can you guarantee the safety of my son if he goes to Northern Ireland?" My answer could only be, "no more than anyone can guarantee one's safety in any country".

Our recommendation was not accepted. The replies to the questionnaire showed that:-

• Competitors from seven countries would not plough.

• Competitors from seven countries were doubtful or very doubtful.

• Competitors from only two countries were agreeable to participate.

• Fourteen WPO Board Members were against holding the Contest in Northern Ireland.

- Only three WPO Board Members were agreeable to hold the Contest in Northern Ireland.

- Thirteen WPO Board Members said they would not go to Northern Ireland.

- Only four WPO Board Members were agreeable to go to Northern Ireland.

- Four WPO Board Members and four WPO affiliates did not reply to the questionnaire.

With short notice the Swedish host organisation, Jordbrukare Ungdomens Forbund, went into action with only two years in which to prepare for the Contest near Vara in the year 1976. Meanwhile, the turn of the Northern Ireland Ploughing Association to host the Contest was postponed until the year 1979.

Second invitation accepted.

At the second attempt three years later the WPO Board, all the affiliates and the ploughmen expressed no qualms about participating in Northern Ireland. The three hundred and fifty acres of polder land reclaimed from Lough Foyle more than a century before was again made available by the generous owner, Mr James Gilfillan. The new University of Coleraine became headquarters 'home' of the world ploughmen. As always in Ireland, a large and comprehensive trade exhibition and machinery demonstration area was included in the plan requiring the generous allotment of extra space from neighbouring farmers. Irish folk from north to south understand the importance of farming and they enjoy to attend ploughing matches. Ploughing begins the farming year.

Westwards, beyond Ireland lies three thousand miles of heaving ocean and, as the weather men remind us, a procession of troughs of low pressure, banks of cloud and impelling winds so often drive the accompanying rain from its vertical fall to the horizontal. For this experience Irish ploughing enthusiasts are always equipped. So, too, must be the world ploughing competitors with rubber boots and rainwear.

The weather proved to be a repeat of what we experienced in County Wexford in the year 1973. Torrential gale-driven rain and thousands of feet churned headlands into a slimy sea of mud. What we did not wish to experience again in addition to the weather was the persistence of the political stirrers. The same principal stirrer who raised controversy when the World Contest was at Wexford, and whom we were told was a foreign gentleman based at Dublin University, wrote to the Northern Ireland Ploughing Association complaining against the presence of two champion ploughmen from Rhodesia. Simultaneously, there came a similar letter from an individual in Germany and a telegram from a person in New Zealand, suspiciously promoted from the same source.

The palaver among politicians over the problems in Rhodesia had not reached a settlement. Fortunately, however, the World Ploughing Contest managed to avoid being adopted again as the platform for controversial political press headlines. Instead, the headlines boldly emphasised ploughing – the purpose of the international ploughmen's presence in Northern Ireland. Unknown to press, public and ploughmen sinister happenings were in the background.

The Contest was held on the 20th and 21st September, 1979. During the night on the eve of the first day two things happened. Outside the west wind blew and rose into an Atlantic storm. Inside the University building shortly before midnight our Hungarian Board Member Pal Vigh, one-time Agricultural Attaché in Washington, USA, and Jiri Donth our Board Member from Czechoslovakia, came to my bedroom. The Hungarian Embassy in London had been informed that two Rhodesian ploughmen were entered in the World Ploughing Contest and, therefore, the event must be boycotted by the ploughmen from countries which applied sanctions against Rhodesia. The fact that ploughmen from Rhodesia were entered in the Contest was not unknown to authorities in all the countries from which the competitors came. Pal Vigh had received a message ordering the Hungarian ploughmen to withdraw from the Contest and return home. He was in great distress but had not told his ploughmen. He had confided the unexpected news to Jiri Donth who had not received any message from the Czechoslovak Embassy but felt that, being from the Communist bloc, one would surely come. The two men were genuinely embarrassed and sorry for their ploughmen who were sleeping peacefully following final practice in readiness for stubble ploughing in the morning.

To withdraw and return home on the first day of the Contest was against their good conscience. They appreciated the sincerity of the welcomes received and the warmth of Irish hospitality. They were in a dilemma, ordered by a dictatorship to suddenly depart and faced with the discourtesy of flouting the loan of ploughs and tractors, costs involved in securing their participation, and the hospitality and fellowship they were enjoying. What would happen if they refused to obey the order? Their grave answer, "The consequences could be serious".

Diplomatic Approach

Eventually, we decided the General Secretary speak by telephone to the Hungarian Charge de Affaires in London, but first all three go together to acquaint the WPO Rhodesian Board Member, Alec Philp, of the situation, since both the Hungarian and Czech Board Members wished to assure him they were in a plight and were not themselves hostile to fellow ploughmen from anywhere. Alec was fast asleep when we knocked on his bedroom door. We all four discussed for about half an hour and firmly decided not to inform the ploughmen, the Board Members, the press or anyone. Meantime, I would telephone the Charge de Affaires and ask him to have the Hungarian order rescinded.

The telephones at Coleraine University were not very privately situated in corridors. Jiri Donth returned to bed. Frank Espley, Secretary of NIPA trustingly lent his motor car to Pal Vigh and myself at one o' clock in the morning, for what reason he never asked and probably never knew until many moons later that it was to find the privacy of a public telephone kiosk in Coleraine town! Why we failed to find one, we never knew!

By the time we returned to the University building everyone had gone to bed so we telephoned the Charge de Affaires office from one of the telephones in a corridor. We felt sure the diplomat would be in bed but cared not. The call was answered by a janitor whom I persuaded to give me the diplomat's home number in London. We then decided to go to bed and Pal would come to my room at 6am when we would call again on the same telephone to Dr Simonyi before he had breakfast. Unless the diplomat was a farmer, to be called so early might not contribute to his pleasure, but we hoped the choice of hour would be psychological, when the mind would be fresh! We gathered as many coins as we could find to feed the telephone. Pal Vigh fed them into the coin box one after the other whilst I did the talking.

Dr Simonyi said the order came from Budapest. "Then please have the order cancelled". That he could not do. Then he was told all the reasons why he should. The action of withdrawal was extremely unpopular and very bad publicity for Hungary, being a gross discourtesy to the hospitable Irish hosts, to Ireland and Irish people generally. Each competitor was being hosted to a cost of about eight hundred pounds, provided with tractors, ploughs and transport, accommodation and subsistence and the fellowship and kindness of friendly people.

As each point of argument was pressed into the telephone Pal smilingly nodded approval and kept pressing coins into the slot. Explained to Dr Simonyi was the fact that all the competitors in the World Ploughing Contest participate voluntarily as individual ploughmen, not representing political parties and not as teams. To withdraw the Hungarian ploughmen would be regarded as discrimination as well as bad manners, and the anger of the disappointed people who had warmly welcomed them in their midst would be severe. More nods and smiles of agreement as Pal plopped coin after coin through the slot! The diplomat was informed that we understood the situation was brought about by someone in Dublin, not an Irish person, who specialised in causing trouble for a political purpose and was taking advantage of this international, non-political, educational and sporting event to gain attention. Dr Simonyi re-affirmed he could do nothing about it.

Then he had to be told about the large exhibition of agricultural machinery and equipment which included some Hungarian products displayed for many thousands of visitors to examine and for some, no doubt, to purchase. The machinery from eastern Europe would be worth many thousands of pounds in value. And then, a shocking hint that property is not always safe when some disappointed people become angry. Perhaps an unfair punch, but Pal was beginning to run short of coins! The explosive reply was that such an idea was preposterous. Hopefully yes, but recalling the reputation the rest of the world had of Northern Ireland when for fear of violence they declined to support holding the Contest there in the year 1976, it was only a hint that some people there do cause explosions!

Pal Vigh's coins were getting fewer and we had not the answer we wanted from the Charge de Affaires. He was told of the embarrassment and the difficulty of returning the tractors and of transporting the ploughmen away when everybody was busily engaged in the organisation. Since the order came from Budapest I begged him to refer back to his Government what he had been told

Sir Robin Kinahan

and request cancellation of the order. For our part nothing would be done until he called back with whatever response he received from Budapest. If he did not reply within the next few hours that same day we would assume the order had been rescinded.

Only four people knew of the development. Only Pal knew of my talk with the Hungarian diplomat. The diplomat did not know that Pal was keeping the line open by dropping coins into the telephone box!

Secrecy

The first day stubble land programme began on time in a downpour of rain and strong wind. Weather conditions threatened to delay the parade to the plots. The Opening Ceremony was to be performed by the President of Northern Ireland Ploughing Association Sir Robin Kinahan, whose interest in competition ploughing dated from the year 1959 when, as Lord Mayor of Belfast, he kindly entertained the world ploughmen of that year in the fine Belfast City Hall. I explained to Pal Vigh we must take Sir Robin into our confidence because he could help to keep things moving whilst performing the ceremony. Our aim and hope was to get all the ploughmen working on their plots and unapproachable before any message came back from Budapest via London.

Together with Sir Robin, Pal and I took shelter from rain, wind and people behind some buildings where none could see or hear us. We told him the whole story in strictest confidence. No one else would be told anything thus to ensure that nothing would be leaked to the press. Sir Robin co-operated admirably. He performed the ceremony on time with a short speech. Encouragingly waved on the ploughmen rode from the marshalling yard through ankle-deep mud into gale force driving rain. Pal was anxious, yet glad to see his ploughmen on their plots oblivious of the political storm but very conscious of the weather storm. Throughout the day a reply of some sort was anticipated from Budapest. None came. The first day's ploughing was over and all the ploughmen returned to the University to wash and change. After dining together they drew their plot numbers for the next day's grassland ploughing, then enjoyed an evening of fellowship – all, including Rhodesians and Hungarians together. Meanwhile, Jiri Donth felt relieved, so far there was no dreaded message from Czechoslovakia.

During the night the gales blew stronger and next morning the exhibition area was a wreck. Tents, marquees, huts, signboards and flags were flattened to the ground. Replacement and re-erection was a major and intensive undertaking. Fortunately, the storm abated and the rain stopped. A press reporter looking for a story said he must write something special. What might he know? He wanted to know the cost of the storm damage. A pleasing choice of subject but for an answer he would have to contact the Show Director. His next question was, 'Did the international ploughmen come out at dawn with tractors to help restore the exhibition?'

"No".

"Why not?"

"First because they did not know of the damage until they arrived on site from the University and, secondly, because they have a fully committed Contest timetable".

"Well, it would make a good story and I'd like to quote a comment from you. Couldn't you say that some of the international ploughmen came to the rescue?"

"No It would not be true".

What exaggeration he would have made of the political problem one shudders to think.

The second day of ploughing grassland proceeded normally according to plan. The Cairn of Peace was unveiled and dedicated and the Ploughmen's Festival Evening for the presentation of awards was an enjoyable and happy function. Pal Vigh felt relieved that a return message had not been received. Jiri Donth was glad that he did not get any message at all. The new World Champion Ploughman was Robert Wieser from Austria, and the Runner-up was Josef Liszt also from Austria. The post-Contest tour ended with a farewell party in Belfast and all the ploughmen went away knowing nothing of the drama in which the Hungarian and Rhodesian competitors were the principals.

There was a sequel. I heard no news from Pal Vigh for a period of about three months when he wrote to apologise for not having kept in touch, expressing good wishes for Christmas, and confiding he had been the victim of some form of penalty or punishment. He died of illness four years later and was succeeded on the WPO Board by Professor Istvan Meszaros.

Chapter 23

RANCOUR OVERCOME, "WALL DEMOLISHED CAIRNES VANDALISED"

Fortunately, behind-the-scenes excitement during the Contest in Northern Ireland was successfully kept secret. Thankfully, political activists found nothing more to be active about so far as international ploughing was concerned. Rhodesia became Zimbabwe and the WPO affiliate changed name from Ploughing Association of Rhodesia to National Ploughing Association of Zimbabwe, and carried on as before.

When the Contest returned to Zimbabwe in year 1983 the welcome was as warm as on the earlier occasion in 1968. Just as did Prime Minister Ian Smith welcome the event during his term of office so, too, did new Prime Minister Robert Mugabe. The same connoisseurs of ploughmanship and conservation involved with the earlier Contest were similarly involved a second time. The erstwhile demand by some impetuous Scandinavians for the expulsion from WPO of Rhodesian-cum-Zimbabweans, whose hospitality they now enjoyed, melted into warm friendship and co-operation. The political rancour was overcome, and need never have been. Such is the fellowship of the ploughing fraternity and the reason why the plough is the traditional symbol of peaceful endeavour.

When the "Wall" came down

At the end of the decade which was the 1980s the infamous dividing wall from north to south across Europe was demolished by the people it had imprisoned. In historical retrospect this was to be ultimately expected. Throughout time frontiers have been defined and changed, erected and eradicated, including Roman walls, Ziegfield and Maginot military lines. Some built to keep people in and others to keep people out.

Following a visit to the original wire "wall" in October 1964 I wrote, "Few people appreciate what an ugly piece of work is the so-called "Iron Curtain" which divides Europe from north to south into an eastern bloc and a western bloc – unless they have seen it. In due course of time this man-made barrier will crumble and rot and be trampled underfoot just as has happened to all walls in history which were erected to satisfy the ego of oppressive political factions".

Since when, of course, the deadly barrier has indeed been crushed to dust out of recognition. The east and west German ploughmen are no longer kept apart. Release came, also, to ploughing organisations in Estonia and Latvia from where, for the first time, their champion ploughmen journeyed freely to Norway to participate at Klepp in the 36th World Ploughing Contest in year 1989. They crossed hitherto restricted frontiers and found a friendly welcome among people of many nationalities from whom for forty years political ideology had kept them ruthlessly isolated.

In the days of horse ploughing we were told a good ploughman having put his hand to the plough does not look back. Paradoxically, the modern ploughman does not follow the plough and, therefore, needs often to look back! In composing this narrative I, too, have looked back over more than half a century of ploughing in competitions and on farms around the world and have come to believe that, since time immemorial, an inherent quality of the rural way life is fellowship, active, potential and latent, often expressed in both verse and prose in many languages and now shared in reality internationally. With the facilities of modern travel, world-wide adoption of the agricultural tractor, the reduction of frontiers and the introduction of world competition ploughing the potential is realised for many in person to person and family to family acquaintanceship.

Looking back also recalls unexpected political hostility attempted to rupture the good example of international amiability that is the desire of every peace-loving world citizen. Political interference unnecessarily prevented and bitterly disappointed no fewer than forty-eight ploughmen, who had accomplished national championship stature from realizing their once-in-a-lifetime entitlement to participate in a World Ploughing Contest and the associated educational and social activities.

Although the unfortunate competitors thus politically victimised were embarrassed and afraid to fraternize, the political interference failed to quell the inherent fellowship of the ploughing fraternity whose common purpose abides in dedication to husbandry of land.

Cairn of Peace

A monument is erected at the place where each World Contest is held and is appropriately called a "Cairn of Peace". Each Cairn is designed locally and incorporates a block of stone from each competitor's homeland engraved with his country's name in his own language.

It was distressing to discover that miscreants damaged two of the Cairns. In year 1963 at Caledon, Ontario, Canada, the Contest was commemorated by a large globe-shaped monument erected in a prominent location near the offices of the regional Council. The engraved blocks naming the participating countries were neatly arranged around the WPO symbol. Some years later whilst re-visiting Caledon with Canadian Board Member Alec McKinney, it was a surprise to see the engraved stones had been replaced by similar size bronze plaques. On enquiring why, it was shocking to learn that some vandal had used the Cairn as a target for rifle practice and shot out every stone!

The second deplorable incident occurred at Shillingford in Oxfordshire, England, in year 1991. The Cairn commemorating the 4th World Contest, held there in 1956, is built of Cotswold stone with the usual engraved blocks inset and was surmounted by a model of the "Golden Plough" trophy fashioned by a blacksmith in Canada and donated by the Canadian Council of Ploughing Associations. After standing safe and sound in the village for thirty-five years local residents were shocked one morning to see the unexpected — the model plough had disappeared! It had been wrenched from its moorings and has never been found.

Fortunately, the Cairn has been restored to glory thanks to the generosity of the Society of Ploughmen who secured the skilled craftsmanship of blacksmith John Greenslade who built and installed a fine identical replacement model plough.

The ill fated Cairn in Canada

The restored Cairn in England and Blacksmith John Greenslade

144

Chapter 24

SECRECY IN LEIPZIG

"When in Rome do as the Romans do" is all right if being in Rome, or elsewhere, is conducive to one's compatibility. But, to travel in lands where cultures are different from one's own and where political ideology is of the dictatorial variety then a wise traveller's precept is to listen intently, see clearly, and speak carefully. The first of three tenets of a Yorkshire motto is commendable; "hear all, see all, say nowt", provided you are not there for the business of discussing, negotiating, and co-operating when it is essential to use words carefully.

Introductions into the east European ploughing fraternity inevitably included meeting political persons. Unless an introduction was to a government person it took a little time to guess who among the assembled company was of political "leiter" status.

Following a visit to a long-established agricultural machinery factory formerly owned by the family Sach, but appropriated by the State to become known by the initials VEB (meaning 'Nationally owned Enterprise') the three senior department directors invited WPO Chairman Walter Feuerlein and myself to dinner. The meeting place was a private room in a small Hungarian restaurant in Leipzig and our host was the VEB Executive Director.

We examined ploughs and other agricultural implements in the factory, and although production was so slow that state farms and co-operatives had several months to wait for delivery of their orders, the innovation awards scheme to encourage employees to introduce ideas for improvements, inventing gadgets and new designs was impressive. Apparently, production depended on waiting for definite orders and not in anticipation of projected demand as in a market economy where competition speeds up delivery.

During an interlude of complete privacy whilst walking together across a large field, well out of earshot of anyone else, one of the directors candidly expressed his condemnation of the whole system and explained communist ideology was a failure. This unexpected, voluntary description was enlightening but best not that a guest should comment. He reassuringly emphasised that what he said was the truth and continued his condemnation of aspects of daily life and work. Then "We will meet tonight at dinner. Please do not tell anyone what I have told you, but believe me". He was promised and assured that the formality was to see, listen and say nothing.

We were six for dinner with the three directors and the Chief, or Managing Director, a young man probably in his late thirties, at the head of the table. The meal was typically Hungarian with kebabs. Conversation was about ploughs and ploughing, soil, crops and farm machinery generally. The Chief spoke only German and appeared uninspired with the agricultural subject. The conversation steered clear of politics which, one guessed, was his watching brief both in and out of the factory.

Before being taken over by the state the factory was owned and operated successfully by Sach Engineering Company. Professor Sach and family fled to West Germany where he obtained a post at one of the Universities. He was an authority on the manufacture of ploughs and strongly advised the adoption of the reversible plough for the purpose of achieving a flat seedbed free of riggs and furrows necessary to cope with the introduction of mechanical seeding, planting and harvesting machines. Especially for crops like sugar beet a seedbed of even depth and surface is needed for the machines to both plant at an optimum depth and at harvest uplift at an optimum depth without damaging roots or tubers.

Shortly after writing at length on this subject to WPO we met at the World Contest in Germany in 1978. He was in poor health but still keenly interested in the ploughing. Somehow he had heard of our visit to his old factory and was eager to hear anything we could tell him about it.

Whilst in Leipzig a lone visit to the silent, deserted St Tomas Cathedral was an emotional experience. The authorities had condemned the building to be demolished, but, fortunately, yielded to powerful objections from the populace. Therein is the tomb of John Sebastian Bach, renowned organist, composer of magnificent choral music, who produced more than two hundred cantatas and a family of twenty children.

The silence in the great empty chamber, bereft of its ecclesiastical components, left to the imagination the resonance of the organ accompanying the choir to sing Bach's oratorios. It was sad to think that this bare and empty fountain from where inspiring music was first poured out to the world no longer raised an echo.

Walter was interested, and at seven o'clock next morning he, too, went alone to St. Tomas Cathedral but arrived to find he was not alone. A group of ten or a dozen young people were grouped before the empty alter space. One was reading quietly from the Bible. He stopped upon noticing Walter's quiet entry. There was a short whispered conversation until one member of the group walked down the aisle, greeted Walter, explained they shared a short service before going to their work places, and would he care to join them in a prayer and a hymn? Walter did so, readily, and admired their integrity.

This was one of several incidents that surely illustrate the human 'cement' of common understanding we call fellowship. Despite ideologies based on the theories of political 'isms' human beings whether white, black, brown or yellow have the same natural desires to eat, love, produce and survive in peaceful relationships. When greed goads mankind into militancy neither those who die nor those who survive have won. Succeeding generations suffer and should learn from the folly of their ancestors but they, also, risk being politically bamboozled. Until nations have leaders without guns in their hands, without corruption, antagonism, enmity, without violence, will inborn "soul power" maintain civilized behaviour.

Just as the fellowship of the international ploughing fraternity held firm against activist attempts to derange cordiality by political confrontation, so, too, did the spirit of the group in St Tomas Cathedral hold to the faith that ultimately broke down the wall of their political prison. Although physically divided for forty-five years the populace was to declare again that *Deutschland ist unteilbar* – Germany is indivisible.

Chapter 25

POLITICIANS : PRACTITIONERS
OF PERSUASION

Totalitarianism is the antithesis of democracy, yet a totalitarian regime impressed upon its people they were safeguarded to enjoy democratic freedom. During our visits beyond the "Iron Curtain" we learned of the so-called freedom, and that not all people were persuaded. We made a private visit to the home of an old forester friend of Walter Feuerlein in the vicinity of Leipzig. It was an emotive occasion for they had thought, under the circumstances of partition, they would never see each other again. In his boyhood this forester had been a Boy Scout, visited London and at the Boy Scout Shop bought himself a Baden Powell Boy Scout hat, a precious souvenir which he brought out to show us with pride.

Following in his father's footsteps, his student son intended to be a forester, too, but faced a serious problem. They were a Christian family and the son wished to be confirmed in the faith the same as were his parents. If he was to be confirmed in a religious faith he would find no advancement in his career beyond being a labourer in the forest despite his ambition to study silviculture and forestry management at University.

To overcome this discrimination his father decided to consult with a Bishop who, strange as it may seem, was a member of the Communist party bureaucracy. An obvious case of if you cannot beat them join them but, of course, harbour a well-meaning motive! The Bishop explained either join the Church and be discriminated against or join the Communist party and succeed. In a totalitarian regime there could be no personal advancement without becoming a member of the party. That was the dilemma. So, said the Bishop, the first essential was to become a card-holding member of the party, and then work hard at college to win a place at University in your chosen forestry subject and qualify for your profession. During this time forget not Christian principles and after graduating go quietly to the Bishop who would gladly perform the rites to confirm his membership of the Christian Church. And that is what they planned to do.

An evening visit to a family in Dresden was also revealing of mental stress under the restrictive system. The husband was by profession an optician but also a playwright of some distinction. After supper he brought out his manuscripts of several plays together with the associated publicity and other advertising material. Following an interesting resume of his works he declared most emphatically that he would write no more. His play-writing was finished. Why? Had he no more ideas, no inspiration?

That was not the reason. He was frustrated through having to submit his manuscripts to the "party" moguls who told him where to make references to, for example, "Russian saviours of our freedom", to prefix the names of characters with the title "Comrade", to make frequent reference to the advantages of the Deutsche Demokratische Republik and maintain a communistic theme throughout whatever the subject of the story.

There was a torrential rainstorm whilst my host drove me in his little Trabant car to the railway station to catch a late night train back to Leipzig. The streets were deserted, and whilst out of everybody's earshot it was a good chance to ask, "What would happen if the barricade on the west was breached?" Without hesitation he said, "If that happens people will leave in their thousands". When that time came, so they did!

Hotel Adlon

Before the wars one of the reputedly, sophisticated social centres of Berlin, where mingled aristocracy and wealth, was the Adlon Hotel. During the second World War the Adlon and surrounding buildings, including the one over Hitler's suicide bunker were bombed. The wall built by the Communists stretched around the west side of Berlin as well as through the middle of the city centred on the Brandenburg Tor where it severed the linking of Bismarck Strasse from the west with Unter den Linden from the east. The bombed ruins of the Adlon were a short walk from the Tor in a southerly direction on the east zone side of the wall. The dilapidated entrance was uninviting but I decided to dine at the Adlon!

This once famous dining and dancing place was a shabby, cluttered ruin of a restaurant. There was one waiter and only one customer – me. The waiter apologised for the candle light – not the romantic variety of years gone by – just candle light because there was no electricity, and a menu without hot meals or hot drinks, no tea, no coffee. The fruit juice kept in the refrigerator wasn't cold because the refrigerator was only cold in winter. An electrician had attended to it five times but it still would not work. Whilst awaiting the serving of the 'feast' the quietness of the place was eerie.

The tete-tetes, music, banquets and frolics that once echoed against those broken walls had to be imagined. After a cold meat sandwich and a glass of luke-warm 'limonade' it was a relief to return to the physical comfort of the new, large and uninspiring architecture of the Hotel Berolina where one's only discomfort, at that time, was the thought that in a room occupied by two visitors from the west care must be taken not to criticise the regime lest conversation might be monitored. As mentioned in an earlier chapter "say nowt".

Chapter 26

MYSTERY HOST IN AUERBACH'S KELLAR

In the Spring of 1966 sixteen members of the former British Ploughing Association flew to Amsterdam Airport there to be met by owner-driver Henk Starrenberg with his minibus and box trailer hitched behind. Luggage was loaded into the trailer and Henk drove the group across the Netherlands and western Germany to the Iron Curtain at the Helmstedt-Marienborn entry control point. Destination was the DDR, East Germany, and mission to visit the farming scene in a communist country, State enterprise farms, Agricultural Research Centres, returning via East Berlin, crossing to west Berlin to observe farming in the blockaded city zone, then across the east zone to exit at Helmstedt into the west again and back to Amsterdam and home.

The visit was arranged by the British Ploughing Association in co-operation with the Deutsche Leistungspflügen on the east side and

Henk's Bus and Max Domsch

153

similarly with the Deutsche Pflügerrat on the west side. The visits were made possible thanks to the extremely helpful liaison of agricultural scientist Max Domsch of the east German ploughing fraternity and by WPO Chairman Walter Feuerlein of the west German ploughing fraternity.

Max arranged to await our arrival as near as permitted in the vicinity of the control point at which we would cross the 186 miles long north-south, man-made divider. We drove to the gap in the wire fences to be confronted by an enormously thick steel hydraulic ram blocking the entrance from side to side. East German soldiers, overseen by a Russian officer, made an initial examination of passports and entry papers before operating the ram to slide open, and directing the bus to be parked beside the first of several offices. The ram then slid horizontally across the road behind us and we felt we were "in" with no possibility of backing out!

All passports and entry documents were collected and strict orders given that no one get out of the bus. Eventually, the General secretary was called to the office to further explain the purpose of our visit, the while making a poor attempt at being interpreter. After a long, long delay we were told we were not expected and our letters of introduction were neither explicit nor in order! Oh, where was Max Domsch? They knew not of him. He was expected to be in the vicinity of the border control. They explained the border control strip was five kilometres (3 miles) wide where, along the road, was a second control. Could Max be waiting there? If he was, he would not be allowed to enter the border strip, nor could he be contacted. I suggested they telephone or send a soldier to find out if he was anywhere near the second control point. The answer to this was negative head shaking.

Whilst this interrogation was in session our farmers cooped up in Henk's small bus were becoming restive. Frequent breaks in the interrogation for back-room consultations enabled me to pop back into the bus to report proceedings and lack of progress, and urge our members to sit patiently and hope. But the delay was becoming too long for physical comfort. Legs were stiff and needed to be stretched. Ultimately, irritability erupted. Enough was enough. With bucolic indifference they all stepped out of the bus. Guards immediately gathered around in alarm, commanding in the German language and gesticulating, but failing to herd them back into the bus.

Not only were they out but they also entered the Guard Room through the "Ausgang" (exit) to look around. Pipe-smokers nonchalantly struck matches and drew on their pipes. Commands, directions, requests, hand signs were all to no effect until one farmer said tell them we want to exchange some money! Once that was understood all were more politely led through the "Eingang" (entrance) to the money exchange counter.

During the serious-cum-comical business of changing pounds into marks we were surprised and relieved by the arrival of our good friend Max Domsch with an escort. Without letting us know soldiers had been despatched to find him and bring him in by car. Max soon settled all problems, money was sorted, passports returned, and on we drove to the east part of Berlin that was the capital of the so-called German Democratic Republic where our accommodation was the Hotel Berolina. Awaiting us was Dr Anton Kunze of the DDR Ploughing Organization and a guide interpreter called Erich.

Visits were made to two large socialist agricultural enterprises at Schenkenbug and Dahlen – each a combination of a number of former independent farms amalgamated into a unit of several thousand hectares and operated like a factory. The staff worked as "battalions", as, for example, cattle battalion, pig battalion, tillage battalion. The work force was all departmental. All the machinery was large for work on expanses of land too large to be described as normal fields having no hedgerows and delineated principally by roadways, tracks and streams. Two research stations were visited. At the Paulinaue Grassland and Marshland Research Institute work was in hand to improve the constituency and structure of low fertility sandy soil regarded as barren except for some coniferous trees.

The method was to introduce humus using a technique called amelioration ploughing, by which half the depth of topsoil together with previously spread either lime , or farmyard manure, change place with a similar amount of subsoil. A specially designed two-furrow plough was used to cut one furrow deeper than the other. The topsoil from the shallow furrow was turned into the deep furrow from which the subsoil was lifted into the shallow furrow. A normal plough slices and moves soil but does not mix. Using this duplex method of interchanging the strata of upper soils some mixing occurs, especially during subsequent harrowing. Skill is required to both correctly set the two different mouldboards and their coulters and when making the initial opening furrows to ensure the topsoil lies beneath the subsoil during subsequent rounds.

At the Soil and Plant Protection Institute at Muncheberg the Professors were very receptive to observations and comments. In Great Britain where many farms are under-manned it is normal for a

155

Amelioration Ploughing

husband and wife to manage a substantially sized farm themselves with the help of occasional or seasonal contract help. In contrast on the extensive State farms where large machinery was working our farmers remarked about the number of people following a one-man machine operation for no apparent purpose. This was a problem, conceded the Professors, of how to employ those whom the machine had replaced. To our group of independent farmers nationalised agriculture was unappealing.

Escape!? — and hospitality in Auerbach's Keller

Our hosts included a government officer whose chauffeur followed the guided tour in company with our bus driver Henk. In whispered conversation the chauffeur kept asking Henk about the possibility of our taking him with us to escape to the west. Henk was most alarmed by his proposition and very wisely explained that nowhere within, under the bus or in the luggage trailer was it possible to hide, and that without any documents to verify he was one of us from the west could he possibly pass through the stringent and dangerous East German frontier control. When the time came to leave, Henk drove us away to prepare for our farewell dinner party in Leipzig. The reluctant chauffeur drove away with his big boss, and his own secret longing.

The final evening was spent in Auerbach's Keller – the cellar attributed to have been the setting for the legendary story of Faust, alchemist and magician, featured by several authors in poems and plays, including Goethe's drama "The Tragedy of Faust", and more than one opera version of "The Damnation of Faust" by Berlioz. Faust was supposed to sell his soul to the devil in exchange for power. Others who have done the same throughout history have found the final dividend to be a bad investment!

The walls of this underground dining room were a picture gallery painted with gruesome scenes of Faust's encounter with Mephistopheles. It was a macabre environment for a festive occasion and in no way an encouragement to sell one's soul to communism, but it was a very good dinner. Professor Raue, head of the Agricultural College of Leipzig University, gave an interesting address on the work of the University in relation to agriculture generally. Max Domsch and several others from the research centres were with us but no one knew by whom we were being hosted. Sometime, after all were seated, and Professor Raue had left, our host did arrive and took his seat at the head of the table. He was probably in his late thirties and very affable. Surprisingly, he spoke very good English and, without revealing his name, told that his mother lived in London, that he was educated in Yorkshire and had been in the British army. He occasionally visited his mother which he could do, so long as the British authorities did not know he was the secretary of the "Anglo-German Friendship Society"!

During the course of the meal conversation was devoted to farming and, of course, ploughing. But after the desert, when coffee was served our host declared we had talked enough about agriculture and should have something more general to discuss about other aspects of our respective countries. For instance, people in the British Isles did not hear or read the truth about the Soviet part of Germany. British news media was undemocratic as was the British Government. He invited questions about life and work and politics in the German Democratic Republic.

Whilst he was listing our misfortunes from not living in a Soviet state a message went round the table from ear to ear agreeing that questions be exclusively related to agriculture. This gave our agricultural friends the chance to hold the stage and tell us more of what we wanted to know about farming. We had no desire to be smothered in communist propaganda. The consequence was that our host found little chance to say much. As the evening wore on he was obviously deflated. As often happened, the British Ploughing Association Secretary was called by his peers to express a vote of thanks. The hospitality was well deserving of a sincere expression of appreciation. It was also necessary to refute the assertion that the British public was deprived of news. On city and railway station newspaper stands both English and foreign newspapers were daily available including the small circulation Communist *Daily Worker*, whereas in East Germany the only English newspaper allowed was the *Daily Worker*! There was little else to say other than the usual courtesies to include Max and his agricultural colleagues, and to raise a hearty round of applause for a most interesting and hospitable time. Our head of table host did not linger, bade us farewell and was first to leave.

Easier to get in than get out of!

From Leipzig we journeyed on to the empty streets and megalithic blocks of buildings in east Berlin. Fortunately, Max Domsch and Erich, the interpreter, remained with us to assist in any serious negotiation at the checkpoint where we would cross the last few metres into west Berlin. Sure enough we had a serious problem. Before actually crossing to the "west" we had to surrender our DDR visitor permit because upon crossing the Berlin wall the visit to the east was over and the permit expired. To get out of west Berlin and enter the DDR again in order to drive one hundred miles westward across to the control-point exit into west Germany at Helmstedt, required applying for a new DDR entry permit. Without a new permit we faced the prospect of having to abandon the bus and fly out of the walled-in city, isolated within the east Soviet zone.

There was no problem about entering west Berlin with our British passports but the eastern Control Command could neither extend nor issue a second permit to allow re-entry into the DDR for our homeward journey. However, with the invaluable aid of our two eastern ploughing colleagues there ensued seven hours of applying, explaining, interrogation, discussing, negotiating and pleading until our appeal was passed to higher authority. Since the time limit on our current permit would end in three days we had the choice of going straight to the Helmstedt/Marienburg crossing or enter west Berlin and risk the result of our application which would be signalled to us

Original "Iron Curtain" (1961) – Looking across from west to east

from an east German military post across the wall on the south side of the city. Henk, our driver, and I would stand near a barrier blocking a road that was criss-crossed by the barbed-wire entanglement fences on each side of the ten-metre broad minefield deathstrip that ringed the west city. From this position in the quiet cul-de-sac the DDR guards would see us when we arrived at the appointed time to await a signal from them.

Max and Erich promised to be in a position on the east side where we could see them, and within calling distance of the military post in case we should need their help. We said our farewells to each other, surrendered our permit to the east German Control, and drove into the zigzag inspection channels of "Checkpoint Charley". The little bus and its trailer was searched inside, outside, and underneath with mirrors. (The escape-minded chauffeur at Muncheberg wouldn't have stood a chance!). Walter Feuerlein, had flown in from the west and in company with Dr. Peter Friedheim, President of West Berlin Farmers' Association, greeted us on emergence from Checkpoint Charley.

159

Same place (1991) on west side

Same place (1991) on east side

It was expedient to make our first call at a west Berlin Police station to explain our situation and hear any advice the Police might offer. They understood and gave directions how to get to where we had agreed to stand in sight of the Communist zone guards and Max and Erich. Their advice was to stand firm and well clear of the demarcation line, wait patiently and hope for a signal.

A broad white line from one side of the street to the other indicated the demarcation. Above and parallel with it lay a barrier pole between brick walls blocking each side-walk, wire fences stretched away to right and left. Ten metres further along the road was a second barrier pole with parallel wire fences also stretching away from right and left as part of the boundary ring. Five hundred metres beyond from where we stood, at the far side of what was termed the "Protective Zone" was the DDR military post. We could clearly see the guards had us in the sights of their binoculars. And there, too, as promised, a little further beyond and to the left stood Max and Erich waving to us.

The only person in the vicinity of our waiting place was a solitary West German policeman who warned us to take great care not even to put a foot on the demarcation line, but to stand well back from it lest an itchy finger might press a rifle trigger on the opposite side. The little guide book on safety precautions to take whilst in the zonal border area, warned that "the border police on the east will shoot without warning at any person entering the Control Strip"

At this very place some days earlier, the policeman told us, an east German guard suddenly made a dash down the road through the control strip, and leapt the wall at one end of the barrier pole to fall on the footpath with wounds from a following burst of gunfire. As he tumbled across the white demarcation line the west German policeman hand signalled not to shoot. The defecting soldier had reached friendly territory, was now a wounded prisoner and quickly removed by ambulance.

After about half an hour a guard came forward from the control post to the second barrier. He raised it and with a wave of the hand beckoned us to walk along the road (a virtual gap in the death strip) to come to the control office. Without words, an officer handed us a new permit to re-enter "der Deutsche Demokratische Republic". After expressing our thanks and appreciation we returned to the German Federal Republic retracing our footsteps with unhurried aplomb to maintain composure and dignity.

Having stepped back into Federal territory, well clear of the white demarcation line, where the west German policeman was still on duty, we looked back to our friends Max and Erich, held up our very necessary piece of paper, and waved goodbye to each other for the last time.

Chapter 27

TRANSFORMED FROM
ROCKETING TO ROOTING

West Berlin was a hive of bustling activity catered for with attractively, well-stocked shops along streets flowing with traffic, and at night time brightly illuminated. But, in both parts of the former whole city there was an awful reminder, the rubble of war. Within the wartime boundary of the blockaded zone that was West Berlin was a surprising area of well managed agricultural land; seldom mentioned in the news media until the introduction of the successful series of Berlin "Green Weeks". Peter Zorn, an intensive farmer on the sandy soils fringing the Havel and Tegeler lakes, was our guide to the substantial production of horticultural crops, milk, pork, poultry and eggs in the few hours left before using our new Soviet zone, time-limited entry document. Our re-entry into DDR involved another intensive inspection before being released on the hundred miles restricted and undeviating route to the Iron Curtain exit at Marienborn for Helmstedt and Braunschweig in the west. A final submission to the time taking, intensive search procedure and we were on our way to the nearby British Forces Helmstedt Checkpoint. There with a sigh of relief at sight of the "Union Jack" our minibus load of farmers burst forth into singing "Land of Hope and Glory".

Walter Feuerlein had returned to Braunschweig by air and welcomed our arrival at the Agricultural Research Centre at Volkenrode, where we were introduced to techniques used in examination of soils and soil tillage experiments. The Institute was an exemplification of "swords being turned into ploughshares". During the second World War (1939-1945) the original Institute was established for the production of rocket bombs ("buzz" bombs and V bombs) as a replacement for crewed bomber aircraft. The only visible evidence of this former centre of a nefarious science were two enormously thick portions of a bombproof concrete wall, drunkenly tilted but set fast in defiance of a demolition explosion that blew to smithereens the test-bed tunnel of which they were a portion.

When the war ended common sense ruled that armament factories make ploughs instead of weapons, and Volkenrode ballistic centre was erased. The ploughing fraternity adopted, rebuilt and developed the laboratories and workshops. The land was made tidy and divided into useful sized plots upon which to test tillage implements and their effects on the structure of the soil and to examine crop root development as a result of various tillage processes. Of major importance was the development of the ever-lasting plough to the adoption of the agricultural tractor.

Volkenrode's scientific researchers led by Walter Feuerlein promoted the ploughmen-scientist relationship and established the ploughing match organisation in west Germany. A feature of the Centre was a collection of soil profiles from many parts of the country preserved by a process of lamination. Walter was expert at examining soil profiles. He loved to examine the soil on every farm he visited. Armed with a square spade for the purpose he demonstrated a simple method of soil "analysis" useful to every farmer. Inserting the spade vertically into the ground he carefully dug out a rectangular slice of earth, lifting gently to avoid sliding, crumbling or squeezing. Having managed to retain the definition of the strata it is possible to see the depth of topsoil, and estimate the sand, clay and humus content. Closer examination will reveal porosity or compaction, fibre, worm activity, strength of hard pan and consequent direction of spread of roots. The moisture content may be visible at different layers and the size and shape of soil particles indicate the extent of aeration. For example, lumps of manure that have not decomposed indicate lack of air circulation.

From such observations a farmer can draw conclusions regarding the right depth to plough and whether there is need to subsoil or drain. His analysis will help him decide what cultivation process to adopt to improve soil structure. For example, by spreading farmyard manure or ploughing in a green crop. Such a "spade investigation" accompanied by a chemical analysis will determine whether lime or other constituents are needed.

Such thoughts about the soil he ploughs occupy the mind of the experienced ploughman who understands the how, why and wherefore of ploughmanship. As manager of the soil he is caretaker of the world's most precious asset.

Chapter 28
SIGNED WITH A PLOUGH

When the detested European wall was finally crushed into dust, and it was again possible to plough across where it once stood, the concept of "ploughing Fellowship" overflowed from ploughing match societies in the released countries. In particular Latvian and Estonian ploughing organisations affiliated with WPO. The Hungarian and Czechoslovak societies, already affiliated, re-organised. At long last the Russians decided to affiliate with WPO, after toying with the idea for more than twenty years.

Requests for ploughing rules and information about participating in the World Ploughing Contest came to WPO from several Russian sources including the Ministry of Agriculture in Moscow, USSR embassies in both London and Washington DC, USA and from Agricultural Institutes in Kiev, Odessa and Moskovskaja oblest. All were amply provided with full information. After so many years of delay came a sudden urge with a spate of telephone calls from the "Russian Committee for Soil Cultivation" requesting to enter three ploughmen in the 35th World Ploughing Contest at Amana in Iowa, USA, held in September 1988.

Unfortunately, the request came five months after the closing of entries which are limited to two competitors only, who must be the national champion and his, or her, runner-up. However, the intention was good and the American host committee was more than delighted at the prospect of including Russian ploughmen as their guests to compete with the rest of the world in the peaceful occupation of ploughing American soil. Much more sensible than the formerly much publicised prospect of competing in "star wars". The Cold War was over and Russians were as welcome as all other ploughmen.

Because the entries were too late to be accommodated in the actual world competition a novel, spectacular alternative was mutually agreed. The Russians appreciated the problem and accepted the role of observers preparatory to participation in future World Contests. But, also being eager to plough, and everybody else wanting them to plough, an arrangement was made to share a joint Russian-American "Plowing in Peace" display of fellowship. Werner Gruber and Steve

King (USA) together with L Albotin and E Veski (Russia) were each equipped with tractors and ploughs. Before 200,000 spectators they represented the world-wide ploughing fraternity by simultaneously ploughing their respective country's initials within a nine hundred metre square plot on a gently sloping hillside.

It was an impressive and unique way of signing a gentlemen's' agreement of caring for the most fundamental of all agricultural operations, using the good earth for the parchment and the peaceful plough for the pen.

The following year the Russian Committee for Soil Cultivation affiliated with WPO. Yuri Nitikin was appointed Russian member of the WPO Governing Board and expressed the hope that other Soviet Republic state ploughing societies would also become members of WPO and enter their ploughing champions in the World Contests.

Sad to relate, release from the fetters of Communism brought unexpected problems when friction developed between politicians in newly independent states, even resulting in deadly warfare. Consequently, the spirit of brotherhood was once more frustrated by the behaviour of political protagonists. Will civilisation ever prevail?

American-Russian Joint Endeavour

166

Chapter 29

THAILAND, INDIA AND PAKISTAN

The Food and Agriculture Organisation introduced a light, steel mouldboard plough to replace the primitive wooden model used by those farmers in India whose pulling power was provided by steers or camels. These FAO ploughs were demonstrated at the Agricultural Engineering Institute near New Delhi where they ploughed a shallow furrow of about four inches, as befitted the sun-scorched earth. To plough deep and turn up subsoil would release any vital moisture it might contain.

In form and dimension the new plough resembled the traditional wooden version, and was necessarily quite different from the longer and broader mouldboards fitted to horse ploughs used for deeper ploughing in temperate climate lands. Horse ploughing technique in New Zealand is similar to horse ploughing in the British Isles except that furrows are slightly more shallow, but not so shallow as in India. It was in New Zealand that some well-intentioned folk thought of an altruistic scheme to save horse ploughs from the scrapyard when tractors took over from horses. This was to ship discarded horse ploughs to India as gifts to under-equipped farmers. The intention was to cover the costs of transportation from personal donations. Whether the scheme ever got started is not known but, in respect of ploughs it was not considered to be a good idea in India, where the Director of the Agricultural Engineering Institute, who appreciated the good intention, explained that such heavier, deeper digging ploughs built to be pulled by heavy draught horses would be impracticable in the hands of farmers limited to camel and steer power.

He did say mechanical aid for India's farmers would be most welcome, but well-intentioned donors should please first make careful enquiries through the Ministry of Agriculture to make sure that anything they wish to give is what an Indian farmer really needs. To an uninformed, willing subscriber a plough is a plough, but just as there are horses for courses, and camels for different courses, there are also ploughs that penetrate and ploughs that scratch and, regrettably, the giver's money would be misplaced.

However, India has rich mineral deposits, and vast quantities of iron ore provide the basis of the country's steel industry. Home production of tractors is in the forefront of farm mechanization which is supportive of the practice of ploughmanship. After attending several ploughing matches in European countries and consulting with WPO, Gursharan Singh, of the Agricultural Machinery Association of India, promoted the "All India Ploughing Match" in December 1964. In attendance was WPO Board Member Al Faunce, Chief of the Agricultural Engineering Branch, Land and Water Division FAO of UNO. The organisers could have had no one better to encourage their enterprise. With support from the Ministry of Agriculture the match was held on the campus of Punjab Agricultural University at Ludhiana. Several thousand farmers attended.

A two and a half mile long procession of tractors led the way to the opening ceremony along a route lined with supporters in national costume singing and dancing. The event was reported to have been a rural life display such as had never before been seen and the first farm festival of its kind in India. There were almost two hundred competitors of whom the winner was Shri Pargat Singh, from the nearby village of Loona, using a two-furrow tractor plough.

Upon the success of the match the Agricultural Machinery Association and its offspring the "All-India Ploughing Organization" decided to organise ploughing competitions throughout India. Obviously, a task of some dimension which would take much time to cover the vast sub-continent whose people speak some one hundred and seventy nine languages, and of whom eighty per cent live in villages. Nevertheless, that was the ambition set before Gursharan Singh and his colleagues. To set the seal on the project they held a seminar in conjunction with the ploughing match, posing the question "Can India Survive Without Mechanisation?" Principal speakers were Ministers of the Government, representatives of the Universities, visiting guests from Czechoslovakia and from the Ford Foundation.

In the words of Gursharan, "In a country where there is a food shortage the problem can never be solved until we adopt improved methods of cultivation and educate our farmers to do good ploughing". He tried for many years to raise a prize fund which would pay the travel costs of India's ploughmen to participate in the World Ploughing Contest but, sadly, did not live to achieve it.

Luxury Interlude

During the 1950s years steel plants and factories in India were being financed by British, West German and Russian money. One of the country's principal tractors was the "Escort", built under Massey-

Ferguson licence. The Escort factory product Supervisor did some work as a presenter on India Television. Together we broadcast a programme about ploughing and the role of the World Ploughing Organisation before attending a rather luxurious reception at a large country house. The function was held in colourfully decorated marquees erected on the expansive lawns. Both the mansion and the surrounding trees, which stretched along the circular drive, were hung with coloured lamps. Along the drive came a stream of limousines to unload finely dressed ladies and gentlemen at the stately portal. The invitation was to a foreign mission reception, but whose?

Each arriving guest was announced over a loudspeaker system and simultaneously welcomed by a host and hostess who were not Indians. The words " . . . from Great Britain, representing the World Ploughing Organisation", turned a few heads. (After all, what was the World Ploughing Organisation?). Following the welcome shaking of hands the hostess was overheard to remark to her coterie, "British! how nice of him to come!"

When we reached a glass and a sandwich my companion explained that the function was hosted by the Deutsche Demokratische Republik trade mission to India ...the Communist Government of East Germany.

Later, in the comfort of my room at the Oberoi International Hotel I thought of the poor and hungry souls, whose lot was undoubtedly in the mind of Gursharan Singh when he campaigned "to educate farmers to do good ploughing", whose only sleeping accommodation was where they lay on the sidewalks of the street I had traversed between two islands of luxury. Some were probably migrants from country villages in search of city jobs, not even as well-off as the two snake charmers sitting beside a dusty country road who, for a couple of rupees, would charm their snake to rise majestically erect from being curled up in its wicker basket. Way out beyond the city limits deep, rich soil — dry, but well supportive of weeds, was waiting for a plough.

The Importance of Learning to Plough

There is a controversial aspect to the vast clearance of rain forest where saws, axes and soil shifting machinery expose the earth to make way for the plough. Perhaps, the trees are taken for what they are worth and a barren area left with no intention to replant or help rejuvenate it with the help of the plough. In Thailand intensive forest clearance was hailed as a victory over the malarial mosquitoes which inhabited the

so-called "Forest of Death" by destroying their habitat and making a healthier living space for human beings. But the clearance is reported to have also robbed other long-term residents of their habitat, including tigers, elephants, buffalo and sharing jungle dwellers. One may well ask, does all the surface of the earth have to be transformed, tamed and ordered for mankind's occupation? Though, better, no doubt, that living space be dominated by mankind than by malarial mosquitoes!

Domination by mankind imposes responsibility to maintain a conservation balance that requires a fair sense of adjustment. It needs to be distinct from the destruction of habitat like that performed by the Myas nation who inhabited the Yukatan Peninsula in Guatamala from the third to ninth centuries Anno Domini. Their clearance project lost them their civilisation. Their method of so-called development was to burn down forests and in the clearing plant seeds among the ashes. Lack of humus and the absence of a ploughed tilth gradually resulted in the ashen soil becoming exhausted. Time and again the unfortunate Myas moved on to burn more forest and resettle upon newly cleared land in order to grow food. Eventually, not only did they exhaust the land but also exhausted themselves.

The Maya civilisation was at its peak around AD500 with magnificent buildings and advanced learning in mathematics and astronomy but thoroughly lacking in the skill of ploughmanship. The population became reduced through hunger and was ultimately subjugated by a few enterprising Spanish settlers.

In Thailand the effort has been made to learn to use the plough; for it is the conservation restoring tool where wasteland is to be repaired. A tractor-ploughing match was organised by Pakchong Farmers' Credit Co-operative together with Krasetsart Agricultural University for disc ploughs. Under 50hp tractors with double discs and over 50hp with three and four discs. According to reports of this competition "the need for intensive instruction which competition ploughing demands was evident from rough treatment by tractor drivers who showed little feeling for machinery. Whereas a water buffalo stops when the plough is fouled against a root or a rock the impulsive tractor driver increases power to keep going with the result that he leaves his 3-point linkage plough behind and damaged".

Another criticism was that some drivers (one could not yet call them ploughmen) who had not learned to set a plough correctly used heavy chains to lock discs from sideway movement. Provided they did not plough too deep and the land was free of hazards they did manage to

travel reasonably straight. If, however. there was a difficult patch and the driver pressed on without a backward look, the steering would lock into a crabwise movement which often resulted in snapping of hitch-pin and/or breaking of cross-shaft – because the whole yoke, being out of balance due to improper setting, was inflexible and over-strained.

To avoid these costly incidents ploughing matches were introduced for their educational and training value and adopted under the Columbo Plan at the Cotton Development Centre on the central plain some one hundred and fifty miles north of Bankok. Farmers were given demonstrations and instruction on how to get the best use out of machinery. The basic exercise was "ploughmanship" from setting the plough on the workshop floor to adjusting it in the soil. When correctly set the plough glides as it slices the soil without dragging. From basic instruction the tractor drivers graduated to competition ploughing as the quickest way to adopting technique and the achievement of high standards.

Across the vast landscapes of Pakistan both primitive and modern equipment is used for soil tillage. Support for competition ploughing became evident in year 1958 when President of the Pakistan Power Farmers' Association, Sardar Mohamed Gazanfarrulah Khan, encouraged the adoption of competition ploughing technique on his six thousand acre Isahkel Estate in anticipation of the eventual participation of Pakistan ploughmen in the World Ploughing Contests. Unfortunately, economic and political handicaps hindered progress in this direction despite the apparent enthusiasm of both the Minister of Agriculture and the Agricultural Development Commissioner for what they termed "this Sport of the Soil".

However, a farmer who was also the tractor operations manager of the Ford Dealer Distribution in West Pakistan took up the idea with the American Agriculture Mission to arrange several ploughing matches to coincide with small country fairs, known as "Melas", in the neighbourhood of Lahore. The most common tillage implement at the time was an eleven-tine cultivator requiring little skill. It was not an improvement on the traditional bullock plough but it did do the job quicker provided the tractor -man drove safely! The second most popular tillage tool was the disc plough, whilst the mouldboard tractor plough was hardly used at all.

The Farm Mechanization Association published a report at the Farm Machinery Exhibition at Lahore in year 1970 estimating how many ploughs and tractors were needed in the country. The requirement for

mouldboard ploughs topped the list of all tillage tools. The report also emphasised that training in the skilful operation of tractor-drawn mouldboard ploughs was a pre-requisite for improving agriculture in a country where prevails a rising population and insufficient food production.

Sad to relate, some farmers in Pakistan objected to ploughing because they claimed ploughing made the land unlevel! Such criticism was, of course, a case of an inexpert operator blaming his tools. The plough in control of a skilled ploughman is essential for making and keeping fields level on the flat and surfaces even on the slope. Pakistan farmers are particular about maintaining an even surface across contours on sloping land, and about land forming methods which translocate soil in order to change a surface into a more desirable contour, either as an aid against water erosion or as an irrigation control. It was necessary to demonstrate the technique of tractor mouldboard ploughing to those who were circumspect as to the efficacy of the mould board to manipulate the soil in their traditional way.

Despite interest and encouragement there arose no substantial support in either India or Pakistan to enable their ploughmen to participate in the World Ploughing "Olympics". There was one enterprising non-ploughman who happened to read a copy of the WPO "Bulletin of News and Information" (distributed world-wide at that time) who wished to see the world and improve his income. He began a correspondence with WPO, not particularly about ploughing, but eventually leading up to an incredible request that the General Secretary be so good as to find an English lady who would agree to marry him in England and then part company with him for ever after the wedding ceremony! He was told, very politely, that good men like him were much needed in Pakistan and that WPO was most certainly not in the lady market.

Chapter 30

A PLOUGHING MATCH
THAT WAS OVERRUN

On one occasion only were the plots at a World Ploughing Contest invaded by thousands whilst the ploughmen were ploughing. This happened on the great plain of the Marchfeld, known as Austria's Grain Bowl, when the Contest was held at Fuchsenbigl. The month was September in the year 1964, when the farmland extending for many miles in all directions was newly ploughed except for some four hundred acres of stubble and a hundred and fifty acres of very green grassland reserved for the two-day Contest – stubble ploughing on the first day and grassland ploughing on the second day – the only patch of lush green, so far as the eye could discern, on the whole range of flat, fertile plain.

It was on the second day that the invaders arrived and swarmed over every plot whilst the ploughmen were engaged in the grassland ploughing competition. They ran along the furrows, rampaged across the newly turned soil and alarmed lady spectators on the headlands. Never has there been so overwhelmingly an invasion at any ploughing match. They had arrived some time in advance of the world event, but not until the competitors began to plough on the second day did they reveal themselves. From all points of the compass across the grey-brown plain they foregathered on the single patch of distinctive green sward.

This incredible intrusion caused wonder and amazement. Political intruders they were not! Food on the Marchfeld was in short supply because the stubbles had been gleaned of pickings before the soil was overturned in the process of arable renewal. The ley was a welcome oasis where the only available grass was like manna in a wilderness for . . . ? . . . seething hordes . . . of field mice!

World Contest on the Marchfeld (1964)

174

HISTORICAL NOTE

In my earlier book "Ploughman's Progress" I told the story of ploughing in many of its aspects, and of how a local ploughing match in the county of Cumberland brought into being the National Ploughing Championship in Great Britain which, in turn, provided the launching pad for the World Ploughing Organisation and the World Ploughing Contest hosted in a different country each year.

My narrative dwelt principally upon the historical purpose of ploughing as the most basic of all agricultural labours and explained how the art and skill of ploughmanship derives from the instinctive resolution of ploughmen to accomplish work of classic quality.

Such resolution is evidenced in the sporting-cum-educational tradition inherent in rural folk. They determine standards of proficiency in respective tasks by participating in friendly competition. These involve every specialised occupation in farming and country-life pursuits generally.

During the years of war from 1939 to 1945 there was neither spare energy nor spare time to engage in the traditional ploughing matches and agricultural shows like in peace time. However, shortly after the end of the war agricultural societies revived their activities or where founded anew. The public desire was to resume a peaceful way of life liberated from horror and restrictions as quickly as possible. Motivators who enthusiastically restored country life events were new-era pioneers.

Thus it was on the 23rd October 1945 when a group of fourteen farmers and businessmen foregathered in a baker's shop in Workington and founded the Workington and District Agricultural Society with the object of encouraging "social intercourse among all who live and work by the land and between rural and urban populations". Within ten months the new Society staged their first annual 3-day Agricultural Show and Industry Exhibition representative of all sections of Cumbrian activity associated with farming; large and small livestock, gardening, fur and feather, bee-keeping, crafts, industrial manufacture and even a Baby Show, plus sport. The event was held in the month of August on the picturesque parkland overlooked by the historic Workington Hall, a one-time resting place for Mary, Queen of Scots, on her northern journey.

The choice of venue was evocative of a much earlier Workington Agricultural Society founded by John Christian, (a kinsman of Fletcher Christian of the mutiny on HMS Bounty) who, having married a Curwen family heiress came to reside at Workington Hall, her family home, and adopt her Curwen name.

John Christian Curwen (1756-1820) was Member of Parliament first for the city of Carlisle and later for the county of Cumberland for more than forty years. He founded the first Workington Agricultural Society in year 1806 and it soon became the largest of its kind in the country. The Society's annual shows and the estate's experimental Schoose Farm attracted distinguished visitors and farming experts from many counties. From the experience of this successful enterprise Curwen and a Mr Bates, a cattle breeder and colleague from Halton in Northumberland, suggested to the Secretary of the Board of Agriculture (nowadays Ministry of Agriculture) the establishment of a National Agricultural Show which became the forerunner of the annual Show of the Royal Agricultural Society of England. Meanwhile, Workington Society had a branch at Wigton and another on the Isle of Man.

The original Workington Agricultural Society ceased to exist after Curwen's demise. Until then a thriving commerce and a flourishing agriculture had worked hand-in-hand, but in West Cumberland industrial expansion into coal mining and steel production raised an urban-minded population with less interest in the agricultural function. However, a century and a quarter later, after six year's of war (1939-1945) and food rationing, the populace which had experienced the need to "dig for victory" recognised the soil to be mankind's most precious asset. Food originates on farmland where crops for man and beast flourish on well-managed soil; where a good dairy cow can produce a pint of milk a day for every one of seventy people. When food supplies are short appetites are sharpened and interest is focussed upon the resources and potential of farming. Farmers are popular when people are hungry! In times of scarcity the call is for "Backs to the Land", as it was during food shortage in wartime.

The national mood was in prime time for a return to sanity and to restore normal peaceful activities. More than thirty organisations became actively involved in the membership of the new Agricultural Society. They included the Town Council, Chamber of Commerce, National Farmer's Union, Newton Rigg Agricultural College, small and large livestock societies, gardeners and beekeepers, local industries and art societies. The earlier Society could not have received

more enthusiastic support than this one founded in the year 1945.

The year-round programme encompassed many interests additional to those catered for in the schedule of the annual three-day Show. Sheepdog Trials were held in the Springtime and a Ploughing Match in the Autumn-Winter period. A series of lectures, discussions, film shows and social events filled many winter evenings, whilst farm visits specially arranged for town-based organisations occupied many summer afternoons and evenings.

The Ploughing Match soon waxed from a local affair into a county event attracting not only nation-wide entries but, also, ploughmen from Canada, Sweden and Ireland to whom we had extended invitations. Upon this basis, as detailed in "Ploughman's Progress" was founded the British Ploughing Association and the National Ploughing Championships (now conducted in England by the Society of Ploughmen, in Wales by Cymdeithas Aredig Cymru and Scotland by the Scottish Championship Ploughing Association). In its turn the original British Ploughing Association promoted the founding of the World Ploughing Organisation. Each national association enters two competitors in the World Contest.

Since its foundation in the year 1952 the World Ploughing Organisation has enrolled in membership similar national ploughing associations in thirty-two countries. In consequence, the original concept of social intercourse and peaceful co-operation extends world-wide in the fellowship of ploughmen, their families and rural and urban fraternities. Their fellowship and their travels to the wide-world locations chosen for the World Contests extend beyond far distant horizons, there to fraternize and plough alongside each other in competition which is both sporting and educational.

In the days of horse ploughing there was a close consultative relationship between the farmer who ploughed and the village blacksmith who made the plough. That relationship between maker and user was well nigh lost with the introduction of tractors and the mass production of ploughs for tractors. However, since its inception in the year 1952 the World Ploughing Organisation has not only brought the factory mass-producer into mutually beneficial collaboration, as once experienced at the local smithy, but similarly enrolled the soil scientist into a tripartite working fellowship.

This development resulted from the series of symposiums on the subject of soil management, of which the first was held in conjunction with the third World Ploughing Contest held at Uppsala University, Sweden, in the year 1955. The value of the conferences linked with the

177

practical work employed by ploughmen from dry, wet, tropical, sub-tropical and temperate lands, considerably widened the scope of consultation, understanding and exchange of inventive ideas. The consequence was that in September 1973, these scientists involved in the World Ploughing Conferences formed the International Soil Tillage Research Organisation (ISTRO) which has grown to a world-wide membership in some eighty countries. It is a token of corporate feeling that "ISTRO" has been described as the "daughter" of WPO!

Undoubtedly, the popularity of the World Contest influenced a better understanding of the value of soil tillage practices and encouraged adoption of techniques that have raised standards of economic ploughing world-wide. Equally important has been the happy extension of international family friendships and common endeavour as symbolised for ages by the enduring, peaceful plough.

Having put one's hands to the plough it is proper to plough straight and true, but to reach the end of the furrow may not always be easy. Sometimes under that green sward or crumbly stubble there lies a hazard. It may be a large stone, a patch of mire or a set-fast rock, that will give the ploughman a frustrating jolt. The progress of an organisation of international status is also prone to unexpected hazards. Unfortunately, in this otherwise peaceful world, there are political opportunists ever ready to leap on the stage of any international event to display themselves to media publicity.

From such spoilsport mischief neither Olympic Sports nor World Ploughing Contests have been exempt. In this narrative I recall occasions when the peaceful plough sliced into the political mire, and what happened behind the scenes that did not impair the immutability of the multi-national grass-root fellowship.